Future of Management: Er
Sustainability, Diversity, ar

Proceedings of ICMR-2024

Edited by:

Koustubh Kanti Ray
Bhuwandeep

Future of Management: Embracing Sustainability, Diversity, and Inclusivity

Proceedings of ICMR-2024

Edited by:

Koustubh Kanti Ray

Bhuwandeep

Routledge
Taylor & Francis Group

First published 2024
by Routledge
4 Park Square, Milton Park, Abingdon, Oxon OX14 4RN

and by Routledge
605 Third Avenue, New York, NY 10158

Routledge is an imprint of the Taylor & Francis Group, an informa business

British Library Cataloguing-in-Publication Data
A catalogue record for this book is available from the British Library

ISBN: 9781032916835 (pbk)
ISBN: 9781003564447 (ebk)

DOI: 10.1201/9781003564447

Typeset in Sabon LT Std
by HBK Digital

Contents

About the Editors

Koustubh Kanti Ray

K.K. Ray is a Professor of Finance at the School of Management, KIIT Deemed to be University. In a career spanning more than two decades, he has served in leading national-level institutions in the country – Indian Institute of Forest Management, Bhopal; Indian Institute of Management, Raipur, and Institute for Financial Management and Research, Chennai. Dr. Ray has a passion for teaching and taught business valuation, financial engineering, merger & acquisition, and corporate finance at one or more of the above institutions. He regularly conducts management training programs for corporate and government executives in finance and related subjects. His research interests are mainly in the areas of corporate finance and financial markets.

Bhuwandeep

Bhuwandeep is a management professional with over 6 years of experience in Plant Management, Quality Circle assessment, Supplier management, Project management Business Development, and Marketing – across Steel, E-commerce, Textiles, Cement, and Mining businesses. His expertise is in the domains of Sales, Marketing, Business Development, Project planning and execution, and Plant operations. He has worked with Global Indian conglomerates like TATA Group and Aditya Birla Group. He is an alumnus of IIT Kharagpur where he was the recipient of the Merit Cum Means Scholarship. He has also done his PGDBM from IIM Bangalore in the areas of Marketing and Sales. He has also done specialization in Marketing from Rennes School of Business, Rennes, France as an exchange program.

List of Figures

List of Tables

Preface

In response to unparalleled challenges and opportunities, the scope of management is undergoing a profound transformation. Organisations must adapt and innovate in order to flourish in an era characterised by rapid technological advancements, climate change, shifting demographics, and evolving social norms. The three pillars of modern management—sustainability, diversity, and inclusivity—reflect a comprehensive approach that prioritises the well-being of people and the planet over short-term profits and reflects a commitment to social responsibility. In the current era of management, sustainability has emerged as a critical issue. Organisations must incorporate ethical considerations into their decision-making processes, reduce their carbon footprints, and implement eco-conscious practices as the effects of climate change become more severe. According to Paul Polman, the former CEO of Unilever, "Sustainability is not a charity; it is a business case." The increasing support for environmentally responsible businesses from consumers, investors, and regulators has made it both a moral and strategic imperative to embrace sustainability. Financial instability is a potential consequence of neglecting sustainability, in addition to reputational harm. Inclusion and diversity are equally essential for the future of management. Globalisation has led to a workforce that is exceedingly diverse, necessitating leaders who prioritise and actively advocate for inclusivity. As Maya Angelou correctly pointed out, "We all should know that diversity makes for a rich tapestry, and we must understand that all the threads of the tapestry are equal in value, no matter their colour." Diverse teams foster innovation and adaptability by bringing a variety of perspectives, while inclusivity guarantees equal opportunities for all individuals. Businesses that disregard diversity and inclusivity are at risk of becoming disconnected from societal progress and becoming uncompetitive. Verna Myers stated, "Inclusion is being requested to dance; diversity is being invited to the party." Organisations can establish a more equitable and fair society and attain sustainable competitive advantages by incorporating sustainability, diversity, and inclusivity.

In acknowledgement of the significance of these concepts, the School of Management at KIIT University has organised an international conference on the theme "*Future of Management: Embracing Sustainability, Diversity, and Inclusivity*." This academic forum seeks to examine and advance the fundamental principles of contemporary management, directing organisations towards a more successful and fair future. The School of Management, KIIT University, Bhubaneswar, hosted the International Conference on Management Research (ICMR-2024) on February 22–23, 2024. The conference, which lasted for two days, had substantial engagement from scholars, educators, and industry practitioners. Around 100 research papers were presented, with a specific focus on the fundamental values of sustainability, diversity, and inclusivity. These presentations were divided among ten simultaneous sessions. Following a comprehensive screening and double-blind review process, the selection committee has approved 15 articles for publication in these proceedings.

This conference has been an exceptional voyage, a culmination of commitment, cooperation, involving numerous individuals. We express our deep appreciation to the KIIT University management for their unwavering technical and financial assistance. We extend our sincere gratitude to the founder, Prof. (Dr.) Achyutananda Samanta, for his constant support and motivation. We express our gratitude to Vice-Chancellor Prof. Saranjit

Singh, former Pro Vice-Chancellor Prof. Tapan Panda, Registrar KIIT University Prof. J. R. Mohanty, and Dean, School of Management Prof. Sumita Mishra, for their collaboration, direction, and assistance in coordinating the international conference. We would like to express our sincere gratitude to the conference advisors for their invaluable assistance and insightful recommendations, which greatly contributed to the conference's resounding success. We would like to express our gratitude to all the members of the ICMR-2024 programme and organising committee, as well as the conference executives and colleagues from the School of Management at KIIT University, for their unwavering dedication in organising the event.

We express our gratitude to the distinguished external guests who participated in and contributed to the keynote session, panel discussion, and valedictory session. We express our gratitude to all the individuals who presented papers and made contributions to the conference. We express our deep gratitude to the reviewers who diligently assessed the research articles and offered valuable feedback, greatly improving the quality of the chapters in this proceeding.

We would like to extend our heartfelt gratitude to the entire team at Taylor and Francis, especially Ms. Shatakshi Mishra, for their efforts in ensuring a seamless and triumphant publication of the ICMR-2024 proceedings.

Bhubaneswar, India — Koustubh Kanti Ray
Bhubaneswar, India — Bhuwandeep

1 Cross-country Connectedness of U.S. Sustainable and other Thematic Indices on Indian and Chinese Equities

Koustubh Kanti Ray[1,a], *Chandrabhanu Das*[2,b] *and Komapalli Sasi Kumar*[3,c]

[1]Professor, School of Management, KIIT Deemed to be University, Bhubaneswar, India

[2]Assistant Professor, School of Business, GITAM Deemed to be University, Hyderabad, India

[3]Associate Professor, School of Business, GITAM Deemed to be University, Hyderabad, India

Abstract

Cross-country connectedness in financial markets has reduced opportunity through diversification for global investors. In the context of this global risk, the study investigates the impact of U.S. thematic indices on Indian and Chinese equities. The study seeks to examine whether there are any cross-country influences from US thematic indices as alternative investments to major emerging markets. Cross-country connectedness is explored in two dimensions: contagion and cross-country risks. S&P REIT Index, S&P Bitcoin Index, and S&P Green Bond Index (a proxy for sustainable index) were selected as US thematic indices. Shanghai Stock Exchange and BSE Stock Exchange were considered for Chinese and Indian equities. The weekly returns between January 1, 2015, and December 31st, 2023, were analyzed for contagion existence and cross-country risk as dimensions of cross-country effect. The data was analyzed through a wavelet coherence framework to detect contagion and Diebold Yimaz spillover index to identify cross-country risk. The results show a weak linkage of green bonds and bitcoin indices in Indian and Chinese equities with much weaker connectedness for Chinese equities. Real estate investment was found to have a substantial influence on Indian equities. The study will benefit investors in both Asian markets to diversify their portfolios and reduce risk through alternative investments. Further, it grabs policymakers' attention to understand the investment patterns in Chinese and Indian equities.

Keywords: Cross-country connectedness, sustainable index, diversification, spillover, wavelet analysis

Introduction

Technological innovations and financial liberalization have integrated international financial markets. This has benefited investors by decreasing transaction costs and providing more investment avenues. The cross-market connectedness has increased the risk exposure of portfolios with reduced opportunities for diversification. Portfolio managers must look for alternative investments in financial instruments to hedge the risk exposure (Mensi et al., 2019). While the world is divided into emerging and developed economies, the U.S. is one of the largest developed economies, and a crisis in U.S. financial markets may transmit to other emerging economies. The impact becomes significant when it originates from the world's leading economies. U.S. market turbulence may act as a channel of contagion and has a crucial role in international equity markets such as India and China. The financial crisis of 2008 originated in the U.S. and transmitted to other nations, swiftly transforming into a global financial crisis. With

[a]kkray@ksom.ac.in, [b]cdas@gitam.edu, [c]skompall@gitam.edu

DOI: 10.1201/9781003564447-1

the U.S. financial market being more developed, investors may seek to create portfolios with innovative financial instruments of the U.S. market. Our study identifies three alternative investments in U.S. financial markets as U.S. thematic indices based on prior studies of investor preference. The idea behind choosing such thematic indices is that their underlying theme is an innovative asset for alternative investment. The first index is based on real estate investment trusts known as REITs, which are tradable real estate investments based on underlying real commercial property.

Most studies in portfolio allocation focus on stocks and bonds. The study in the REIT space is limited, and there needs to be more concentration in finance literature. Recently, REITs have emerged as an alternative investment vehicle that provides diversification benefits (Monteiro et al., 2024). Historically, they have been attractive investments for improving risk-return tradeoffs in portfolio creation (Bhuyan et al., 2014). The second thematic index is based on Bitcoin, a popular virtual currency. Bitcoin is considered an alternative currency due to its freedom from government intervention. Recent studies suggest bitcoin as an investment asset (Maghyereh and Abdoh, 2020; Corbet et al., 2020).

The final thematic index underlying asset is green bonds. Green bonds primarily finance clean environment projects but have risen to prominence recently. Retail and institutional investors prefer it as a risk management tool (Haq et al., 2021; Saeed et al., 2020). The study intends to find whether there are any cross-country influences from these alternative investments to major emerging markets. China and India are the second-largest economies on the basis of their market size (Li et al., 2020). Chinese and Indian economies are state-controlled capital markets, and stock prices are influenced by domestic policy and foreign investment (Karolyi and Stulz, 2003). Therefore, crises in U.S. markets may serve as channels of contagion and play a crucial role in international equity markets such as India and China. Hence, Chinese and Indian investors must be careful about turbulence in the U.S. market before investing in U.S. financial instruments. Our study is based on this background to identify any scope for diversification in U.S. alternative investments for Indian and Chinese investors. Here, the Shanghai index is considered for the Chinese equity market, and the BSE Sensex is represented by Indian equities. The study has implications for portfolio risk management as weaker or zero linkages indicate benefits for portfolio diversification through a lesser possibility of risk transmission. Findings from this study will be helpful for portfolio managers to devise their risk-return strategies accordingly. Policymakers can benefit by understanding the investment patterns of Indian and Chinese investors.

Literature Review

Cross-country effects can take the form of two dimensions: contagion and cross-country risks. Contagion and cross-country risks were experienced during the financial and currency crises with the integration of global financial markets. Seth and Panda (2018) refer to contagion as cross-country shocks that cannot be explained through fundamental channels. Extremely volatile stock markets may be the reason for uncertainty in policy decisions (Antonakakis et al., 2013). Prior studies can be segmented into two strands.

The first strand is related to studies on financial contagion. The contribution to financial contagion studies has been growing in recent years. Dewandaru et al. (2016) and Yang et al. (2016) have implemented wavelet analysis to study the existence of contagion in Asian financial markets. The study method by Yang et al. (2016) was wavelet coherence analysis. Present empirical works mostly used wavelet coherence to study financial contagion (Jiang and

Yoon, 2020; Kang et al., 2019). The wavelet coherence method eliminates the need to partition samples into subsamples due to its visualization maps, which reveal the nonlinear association between the variables and structural breaks in series (Ali et al., 2020). According to Dahir et al. (2018), wavelet is a better approach than Granger causality or vector auto-regression for examining the contagion effect. Current work by Hung (2021) supports the wavelet coherence framework as a better way to explain contagion simultaneously in multi-scale time and frequency domain. Li et al. (2020) recently evaluated contagion from U.S. economic policy to stock markets in China and India using the wavelet coherence technique. Taking forward from prior studies on the first strand, this study will contribute to the present literature on the contagion of U.S. influence on Asian equities by employing the wavelet coherence analysis.

The second strand of literature focuses on spillovers for cross-country risks. Recent studies report spillovers among asset classes (Farid et al., 2021; Le et al., 2021). The global financial crisis and the recent covid 19 pandemic have lessons for investors and implications for policymakers to explore spillovers across asset classes. The impact of such turbulences has been investigated by the seminal work of Diebold and Yilmaz (2012) to construct a volatility spillover index, which is termed as Diebold and Yilmaz spillover Index that uses generalized variance decomposition from vector autoregression (VAR). Further, it predicts directional volatility spillovers measured in two directions:

a) Spillover received by the market from all other markets
b) Spillover transmitted by a market to all other markets

This seminal work estimates the spillover index that has contributed significantly to evolving studies on cross-country risk. Mensi et al. (2023) implemented the Diebold and Yilmaz spillover Index to examine the spillovers between green bonds, gold, and U.S. equities. Newer studies argue that financial markets analysis by Diebold and Yilmaz spillover Index is preferable to wavelet analysis (Kaya and Demirel, 2024). Habibi and Mohammadi (2022) execute the Diebold and Yilmaz spillover Index and examine the country risk through shocks to the U.S. and its spillover to MENA countries. Higher cross-country risks during a crisis increase portfolio volatility and limit diversification opportunities (Suleman et al., 2023). Investment-related skepticism arises when cross-country risks exist in the portfolio of investors. The variables included in this study have significance in portfolio diversification for global investors. However, prior studies have yet to discuss the nexus between these variables. Hence, this study will be necessary for investors to identify alternative investments in diversifying their risk and generating arbitrage opportunities. Therefore, this study aims to identify the associated hazards in international portfolio diversification through study objectives in the form of research questions as follows:

a) Does there exist contagion from U.S. thematic indices to the two major Asian stock markets
b) Is there a cross-country risk from U.S. thematic indices to the two major Asian stock markets

Methodology

The dataset consists of daily closing prices of all the variables under study from January 2015 to December 2023 based on the availability for all indices. Daily returns are converted to weekly returns estimated as a percentage change in the logarithm of closing prices each Thursday. Returns are calculated in weekly intervals from Thursday to Thursday. Wednesday closing prices are considered if Thursday closing price is not available. This results in a data

Table 1.1 Description of the variables with their abbreviations used in the study and source from which the data is accessed.

Abbreviation	Index	Description	Source
GBOND	S&P Green Bond	Tracks the global green bond market	https://www.spglobal.com/spdji/en/indices/esg/sp-green-bond-index/#overview
BITCOIN	S&P Bitcoin	Tracks the Bitcoin performance	https://www.spglobal.com/spdji/en/indices/digital-assets/sp-bitcoin-index/#overview
REIT	S&P United States REIT	To gauge the publicly traded real investment trusts residing in the U.S.	https://www.spglobal.com/spdji/en/indices/equity/sp-united-states-reit/#overview
SSEI	Shanghai Composite	Consists of A-shares and B-shares traded on the Shanghai Stock Exchange in China	https://www.wsj.com/market-data/quotes/index/CN/XSHG/SHCOMP/historical-prices
BSEEX	BSE Sensex	Consists of 30 well-established companies listed on Bombay Stock Exchange in India	https://www.wsj.com/market-data/quotes/index/IN/1/historical-prices

Source: Authors own computation

set with 416 observations. The U.S. indices were retrieved from S&P Global, whereas the information on the Shanghai Stock Exchange and Bombay Stock Exchange were accessed from the Wall Street Journal website.

Table 1.1 describes the variables used in the study with abbreviations and the source of data. These abbreviations were used to identify the variables throughout the study to conserve space when displaying tables and figures.

Table 1.2 Structure of a connectedness table showing the variance decomposition matrix.

$a1$ $a2$. aN	From others
$a1$ r^h_{11} r^h_{12}	r^h_{1N} b $\sum^N_{j=1} r^h_{1j}, j \neq 1$
$a2$ r^h_{21} r^h_{22}	$.r^h_{2N}$ $\sum^N_{j=1} r^h_{2j}, j \neq 2$
aN r^h_{N1} r^h_{N2}	r^h_{NN} $\sum^N_{j=1} r^h_{Nj}, j \neq N$
To Others $\sum^N_{i=1} r^h_{i1}, i \neq 1$ $\sum^N_{i=1} r^h_{i2}, i \neq 2$ $\sum^N_{i=1} r^h_{iN}, i \neq N$ $\frac{1}{N} \sum^N_{i=1} r^h_{i2}, i \neq j$	

Source: Authors' computation

Wavelet Coherence Analysis

To test the first research question on contagion, the wavelet coherence model is utilized. Wavelet coherence analysis graphically portrays the association between variables by decomposition into several time-frequency components. The co-movements in time series for two-time series variables a(t) and b(t) using cross wavelet transforms (CWT), $W^a_n(u,s)$, and $W^b_n(u,s)$ are depicted in equation (1).

$$W_{a,b}(x,y) = W_a(x,y)\, W^*_b(x,y) \qquad (1)$$

Here W_a (x,y), and W^*_b (x,y) are CWT of time series variables a(t) and b(t). x is the position index, y is the scale, and * exhibits the complex conjugate. The wavelet transform explains the localized covariance between two-time series. The coherence will show regions of varying wavelet power, although co-movement is present.

Table 1.2A Interpretation of various terms from the connectedness matrix.

Representations	Interpretations
$L^h_{i\leftarrow j} = r^h_{ij}$	Pairwise directional connectedness from to
$L^h_{i\leftarrow \bullet} = \sum_{j=1}^{N} r^h_{ij}$, $j \neq i$	Total direct connectedness from others to. This is the h-step forecast error variance of shocks received by from other variables.
$L^h_{i\leftarrow \bullet} = \sum_{j=1}^{N} r^h_{ij}$, $j \neq i$	Total direct connectedness to others from J. This is the h-step forecast error variance of shocks from to other variables.
$L^h = \frac{1}{N}\sum_{i,j=1}^{N} r^h_{ij}$, $j \neq i$	This is the estimation of total connectedness as a single value.

Source: Authors own computation

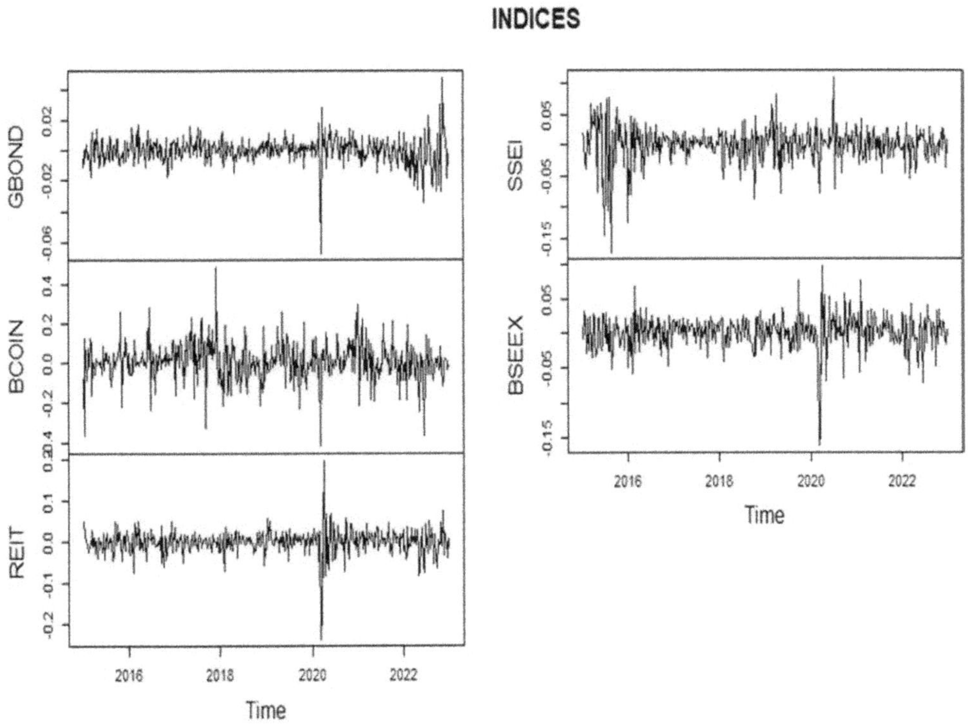

Figure 1.1 Weekly Time Series
Source: Authors' computation

Diebold and Yilmaz Spillover Methodology
The second research question on cross-country risk is assessed using the Diebold and Yilmaz (2012) spillover methodology to explore dynamic connectedness. The spillover is interpreted through a connectedness table.

Table 1.2 gives the schema of a connectedness table. The variance decomposition matrix denoted by $R^h = [r^h_{ij}]$ embeds the variance decompositions. Here, r^h_{ij} explains the part of variable s h-step forecast error variance caused by shocks from

Table 1.3 Summary Statistics for Comparison of Return and Variability.

INDEX	Mean	Median	Max.	Min.	S.D.	Skewness	Kurtosis
GBOND	–0.000035	0.0003	0.0473	–0.067	0.00970	–0.56098	6.3527
BITCOIN	0.01052	0.00525	0.4874	–0.413	0.09650	–0.09114	2.8754
REIT	0.00021	0.00260	0.1963	–0.2376	0.31480	–0.90782	11.7770
SSEI	–0.00022	0.00200	0.1102	–0.1725	0.28194	–1.09667	5.9084
BSEEX	0.0020202	0.00360	0.0975	–0.1601	0.23954	–1.16085	7.9423

Source: Authors own computation

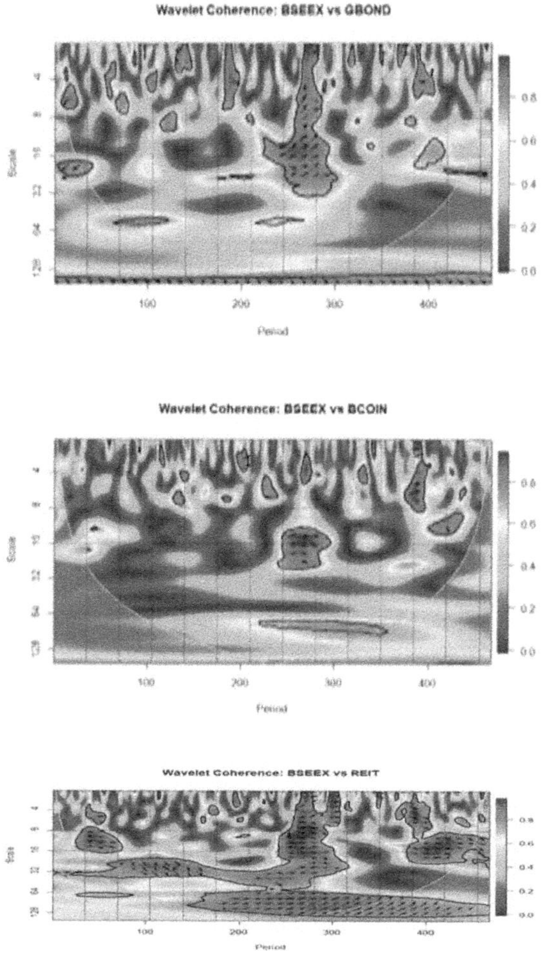

Figure 1.2 Wavelet coherence between BSE Sensex and U.S. Thematic indices. The right side of the chart displays the correlation level; a higher correlation is red, and a blue color depicts a weak correlation

Note: Arrows explain the causality and phase differences. The arrows → and ← suggest the phase and out-of-phase movements. In phase infers a positive association, while out phase means a negative association. The arrows ↗ and imply the lead and lag relationship

Source: Authors' computation

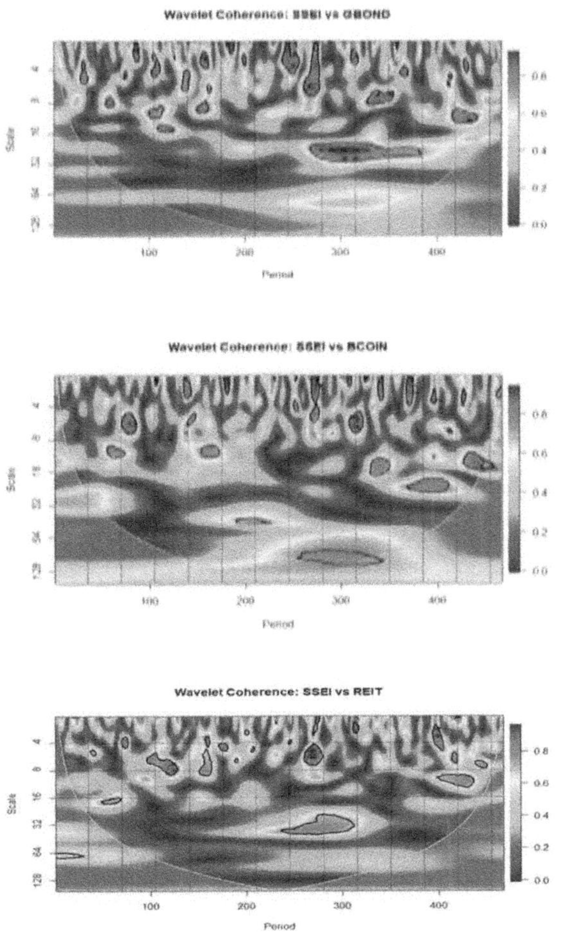

Figure 1.3 Wavelet coherence between the SSE and U.S. thematic indices. The right side of the chart displays the correlation level; a higher correlation is red in color, and a blue color depicts a weak correlation

Note: Arrows explain the causality and phase differences. The arrows → and ← suggest the phase and out-of-phase movements. In phase infers a positive association, while out phase means a negative association. The arrows ↗ and imply the lead and lag relationship

Source: Authors' computation

variable. The off-diagonal elements in the matrix are the specific pairwise directional connectedness.

The connectedness table will give us the following interpretations interpretations as seen in Table 1.2A:

Results

Figure 1.1 shows the weekly return series of variables being discussed in the study. It is observed that all series during the study experienced some volatility clustering. Between 2020 and 2021, due to the outbreak of COVID-19, there was high volatility clustering for all asset classes. It provides evidence of pandemic impact on the returns of these five asset classes.

Table 1.3 shows the summary statistics for weekly returns. The mean return is negative for green bonds (–0.000035) and the Shanghai index (–0.000220). Bitcoin has

Table 1.4 Spillover Index and Connectedness table. The form Indicates Spillover from the Variable, while the form Indicates Spillover to the Variable. To and from others implies spillover to and from the system.

TO/FROM	GBOND	BITCOIN	REIT	SSEI	BSEEX	FROM Others
GBOND	70.13	1.53	15.84	2.48	10.02	29.87
BITCOIN	1.80	89.81	4.77	0.97	2.65	10.19
REIT	13.61	3.26	59.82	3.27	20.05	40.18
SSEI	3.59	0.37	4.39	86.92	4.74	13.08
BSEEX	9.60	2.04	21.70	3.30	63.36	36.64
TO Others	28.59	7.2	46.7	10.01	37.47	129.97
Incl. Own	98.72	97.01	106.51	96.93	100.83	
					Spill over index	(32.49)1.25%

Source: Authors own computation

the highest mean return (0.01052). Hence, it indicates the investor's attention and the popularity of Bitcoin. Based on standard deviation estimates, it is found that the real estate asset class has more variability (0.31480). This supports the opinion of Zhou and Anderson (2013) that herding has been prevalent in U.S. real estate investment assets in recent times, with investors being more responsive to market reactions.

Figure 1.2 displays the intercorrelation between BSE Sensex and U.S. thematic indices. The figure shows that the contentedness for more extended frequency was found only with US REIT compared to other indices. REIT and green bond exhibited an in-phase relationship as seen by right-pointing arrows. Figure 1.3 shows the wavelet coherence between the SSE index and U.S. thematic indices. The color is lesser, unlike BSE Sensex, implying a weaker association.

Although wavelet coherence discussed contagion effects, the visualizations must be more comprehensive to identify cross-country risk transmission between asset classes. Hence, the study's second hypothesis tests cross-country risk transmission and cannot be addressed using only wavelet analysis.

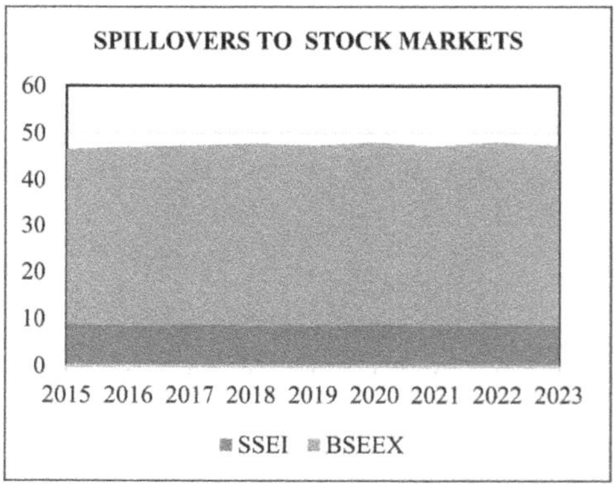

Figure 1.4 Mean Return Spillovers Year wise to Indian and Chinese Stock Markets from U.S. thematic Indices. The blue color indicates BSE Sensex, and the red color indicates the Shanghai Index

Source: Authors' computation

Therefore, the Diebold and Yilmaz (2012) spillover approach was employed to quantify the cross-country risk transmission. Table 1.4 displays the findings of the ten-step-ahead variance decomposition for index returns. The spillover index is 32.49 percent, which suggests that risk transmission in different asset classes cannot be attributed solely to spillovers. Among the two major Asian equities, the Shanghai

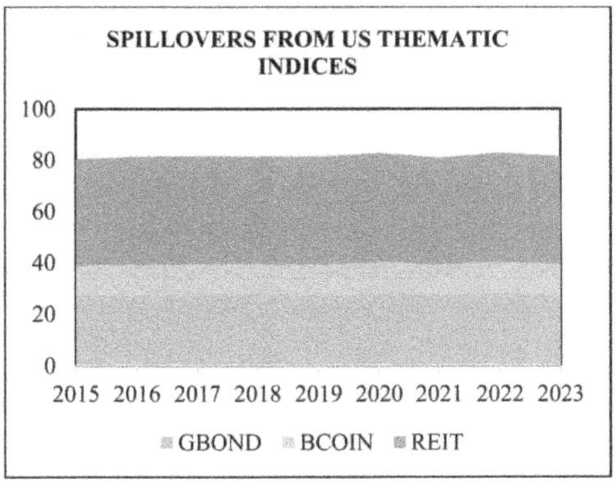

Figure 1.5 Mean Return Spillover Years from U.S. Thematic Indices to Indian and Chinese Stock Markets. Orange Indicates Real Estate Investment, Grey Represents Bitcoin, and Green Indicates the Green Bond Index

Source: Authors' computation

exchange has a higher risk of transmission from domestic shocks (86.92) than the Sensex (63.92).

This reveals that Indian equities are more susceptible to external shocks, unlike Chinese equities, which can be affected by the Country's macroeconomic turbulences. It is seen that cross-country risk from thematic indices to Sensex, although small, is more compared to the Shanghai index, wherein the numbers range from 2.48 due to green bonds to 3.27 from the real estate asset class. U.S. real estate and Sensex have higher spillovers compared to other asset classes. However, bitcoin's contribution to spillover in the two Asian equities is lower than that of other thematic indices. This aspect of Bitcoin makes it a safer alternative investment, in line with Su et al. (2020) and Dyhrberg (2016) studies.

Figure 1.4 plots the mean return spillovers to Asian equities from market and thematic indices. The spillover is lower for the Shanghai exchange than Sensex, providing more opportunities for international portfolio diversification. This evidence contrasts the study by Arora and Vamvakidis (2011), where they argued about international spillovers to China. The results resonate with Fernald and Babson's (2019) view on the resilience of China's financial system to the global crisis.

Figure 1.5 plots the mean spillovers to stock markets from Asian thematic indices. It is seen that spillover from real estate investment is higher, with no scope for diversification benefits. Bitcoin spillover is lower than other thematic indices, encouraging its addition to the international portfolio.

Discussion

Findings

The results are consistent across different approaches for contagion effect and spillover risks. Overall results provide evidence of a weak association of U.S. thematic indices with the Shanghai Stock Exchange. Significant studies about the U.S. financial markets' influence on the Chinese stock market are limited. The findings contribute to a new study by Li and Chen (2021) that discussed global turbulences' trivial impact on Chinese programs. For the BSE index, except the US REIT index, the other U.S. indices indicate short-term influences even though their association is stronger than the Shanghai index. The findings align with the previous study by Ahmad et al. (2005), which mentioned a weak short-term causality between Indian equities and the U.S. stock market.

Studies about the U.S. financial markets' impact on Indian equities in the current context are scarce, with no consensus between the studies. Therefore, the findings are essential to the extant literature on Asian financial markets. The spillover and contagion effect of Bitcoin with the two major Asian stock exchanges was predominantly low. This indicates that the U.S. Bitcoin index in the global investor portfolio provides ample scope for portfolio diversification. Prior studies have addressed

Bitcoin as a speculative investment (Phillip et al., 2018; Sensoy, 2019; Baur et al., 2018; Bouri et al., 2021). The findings are from recent studies discussing the diversification benefit of Bitcoin as a financial instrument (Okorie and Lin, 2020; Gil-Alana et al., 2020; Conlon et al., 2020). The weaker influence of the green bond index in both Asian equities give opportunities for portfolio diversification for socially responsible investors. The inclusion of green bonds in Investor portfolios is supported by growing empirical studies (Reboredo et al., 2020; Nguyen et al., 2021; Ferrer et al., 2021). Contagion and spillover risks of real estate investment are more substantial for the BSE index than the Shanghai index. While it seems that REITs have captured investors' attention, U.S. REITs can also increase the portfolio risk for Indian investors. However, the findings disagree with studies favoring real estate investment (Ling et al., 2020; Zhu and Lizieu, 2022). Studies on REIT have mainly not mentioned U.S. REIT's influence on Asian stock markets. The findings are crucial given that U.S. REIT has a substantial influence on the BSE index.

Limitations

The study is limited to the stock market of only two countries, India and China, major Asian economies. Other Asian markets were not discussed in the study, and their inclusion would have brought perspective on Asian economies. The cross-country risk was measured in terms of only return spillover, while volatility spillover was beyond the scope because of a need for more information on high and low closing price-related information from U.S. thematic indices.

Implications

The study identified the U.S. thematic indices as a proxy for alternative investments on various themes that can interest investors. While the literature on diversification and hedging is advancing rapidly, the study contributes to the literature by finding investment-related uncertainties in financial markets. The dissimilarity in shock transmission between Indian and Chinese equities grabs the attention of policymakers who are trying to devise measures to ensure financial stability. Finally, the study will be practical for institutional investors in optimizing asset diversification and allocation techniques for an investor's portfolio.

Future Directions

The study can be extended to BRIC countries or other country groups to identify investment alternatives in developed financial markets. Further, company-specific studies can be conducted to identify contagion and spillover effects in firms of different economies.

Conclusion

The paper examines the dynamic connectedness between U.S. thematic indices and two major Asian equities, India and China. The dynamic connectedness has been tested in two dimensions. One is the contagion effect, and the other is cross-country risk. The U.S. thematic indices were identified on basis of extant literature about investor preference. Wavelet coherence was employed to find the presence of contagion. Diebold and Yimaz's (2012) spillover index methodology was implemented to explore cross-country risk. The results indicate a weaker association with U.S. thematic indices except for real estate investment, where the results differ for Indian and Chinese equities. Real estate seems to influence Indian equities profoundly and is not advisable for portfolio diversification for Indian investors. The findings are different for bitcoin, which has been under debate as a speculative asset but was found suitable for reducing risks in diversification in the study. The study revealed the resilience of the Chinese economy to external shocks from developed economies.

Based on the findings, investors should have moderate exposure to U.S. alternative investments in their portfolio management. The preference for real estate investment must be evaluated considering the risk transmission to Indian equities. The study contributes significantly to the extant financial risk and portfolio management literature. Finally, policymakers can contemplate the findings from this study to explore the regulatory interventions behind differences in response to shocks between Indian and Chinese economies.

References

Ahmad, K. M., Ashraf, S., and Ahmed, S. (2005). Is the Indian stock market integrated with the U.S. and Japanese markets? An empirical analysis. *South Asia Economic Journal*, 6(2), 193–206. DOI: 10.1177/139156140500600202.

Ali, R., Butt, U. U., Khan, M. M., Shaheer, M., and Zaidi, F. A. (2020). Empirical evidence of co-movement between the Canadian CDS, stock market and TSX 60 volatility index: a wavelet approach. *SEISENSE Journal of Management*, 3(3), 51–64. DOI: https://doi.org/10.33215/sjom.v3i3.353.

Antonakakis, N., Chatziantoniou, I., and Filis, G. (2013). Dynamic co-movements of stock market returns implied volatility and policy uncertainty. *Economics Letters*, 120(1), 87–92. https://doi.org/10.1016/j.econlet.2013.04.004.

Arora, V., and Vamvakidis, A. (2011). China's economic growth: international spillovers. *China and World Economy*, 19(5), 31–46. https://doi.org/10.1111/j.1749-124X.2011.01257.x.

Baur, D. G., Hong, K., and Lee, A. D. (2018). Bitcoin: medium of exchange or speculative assets?. *Journal of International Financial Markets, Institutions and Money*, 54, 177–189. https://doi.org/10.1016/j.intfin.2017.12.004.

Bhuyan, R., Kuhle, J., Ikromov, N., and Chiemeke, C. (2014). Optimal portfolio allocation among REITs, stocks, and long-term bonds: An empirical analysis of U.S. financial markets. *Journal of Mathematical Finance*, 4(2), 9. DOI:10.4236/jmf.2014.42010.

Bouri, E., Gabauer, D., Gupta, R., and Tiwari, A. K. (2021). Volatility connectedness of major cryptocurrencies: the role of investor happiness. *Journal of Behavioral and Experimental Finance*, 30, 100463. 100463.https://doi.org/10.1016/j.jbef.2021.100463.

Conlon, T., Corbet, S., and McGee, R. J. (2020). Are cryptocurrencies a safe haven for equity markets? an international perspective from the COVID-19 pandemic. *Research in International Business and Finance*, 54, 101248. https://doi.org/10.1016/j.ribaf.2020.101248.

Corbet, S., Katsiampa, P., and Lau, C. K. M. (2020). Measuring quantile dependence and testing directional predictability between Bitcoin, altcoins and traditional financial assets. *International Review of Financial Analysis*, 71, 101571. https://doi.org/10.1016/j.irfa.2020.101571.

Dahir, A. M., Mahat, F., Ab Razak, N. H., and Bany-Ariffin, A. N. (2018). Revisiting the dynamic relationship between exchange rates and stock prices in BRICS countries: a wavelet analysis. *Borsa Istanbul Review*, 18(2), 101–113. https://doi.org/10.1016/j.bir.2017.10.001.

Dewandaru, G., Masih, R., and Masih, A. M. M. (2016). Contagion and interdependence across Asia-Pacific equity markets: an analysis based on multi-horizon discrete and continuous wavelet transformations. *International Review of Economics and Finance*, 43, 363–377. https://doi.org/10.1016/j.iref.2016.01.002.

Diebold, F. X., and Yilmaz, K. (2012). Better to give than to receive: predictive directional measurement of volatility spillovers. *International Journal of Forecasting*, 28(1), 57–66. https://doi.org/10.1016/j.ijforecast.2011.02.006.

Dyhrberg, A. H. (2016). Bitcoin, gold and the dollar–A GARCH volatility analysis. *Finance Research Letters*, 16, 85–92. https://doi.org/10.1016/j.frl.2015.10.008.

Farid, S., Kayani, G. M., Naeem, M. A., and Shahzad, S. J. H. (2021). Intraday volatility transmission among precious metals, energy and stocks during the CO-

VID-19 pandemic. *Resources Policy*, 72, 102101. https://doi.org/10.1016/j.resourpol.2021.102101.

Fernald, J. G., and Babson, O. D. (2019). Why has China survived the Asian crisis so well? What risks remain?. In Financial Market Reform in China. Routledge, pp. 55–85. https://www.taylorfrancis.com/chapters/edit/10.4324/9780429037085-4/china-survived-asian-crisis-well-risks-remain-john-fernald-oliver-babson.

Ferrer, R., Shahzad, S. J. H., and Soriano, P. (2021). Are green bonds a different asset class? evidence from time-frequency connectedness analysis. *Journal of Cleaner Production*, 292, 125988. https://doi.org/10.1016/j.jclepro.2021.125988.

Gil-Alana, L. A., Abakah, E. J. A., and Rojo, M. F. R. (2020). Cryptocurrencies and stock market indices. Are they related?. *Research in International Business and Finance*, 51, 101063. https://doi.org/10.1016/j.ribaf.2019.101063.

Habibi, H., and Mohammadi, H. (2022). Return and volatility spillovers across the Western and MENA countries. *The North American Journal of Economics and Finance*, 60, 101642. https://doi.org/10.1016/j.najef.2022.101642.

Haq, I. U., Maneengam, A., Chupradit, S., Suksatan, W., and Huo, C. (2021). Economic policy uncertainty and cryptocurrency market as a risk management avenue: a systematic review. *Risks*, 9(9), 163. https://doi.org/10.3390/risks9090163.

Hung, N. T. (2021). Co-movements between Bitcoin and other asset classes in India. *Journal of Indian Business Research*, 13(2), 270–288. https://doi.org/10.1108/JIBR-03-2020-0071.

Jiang, Z., and Yoon, S. M. (2020). Dynamic co-movement between oil and stock markets in oil-importing and oil-exporting countries: two types of wavelet analysis. *Energy Economics*, 90, 104835. https://doi.org/10.1016/j.eneco.2020.104835.

Kang, S. H., McIver, R. P., and Hernandez, J. A. (2019). Co-movements between Bitcoin and gold: a wavelet coherence analysis. *Physica A: Statistical Mechanics and its Applications*, 536, 120888. https://doi.org/10.1016/j.physa.2019.04.124.

Karolyi, G. A., and Stulz, R. M. (2003). Are financial assets priced locally or globally?. *Handbook of the Economics of Finance*, 1, 975–1020. https://doi.org/10.1016/S1574-0102(03)01025-2.

Kaya, O., and Demirel, M. (2024). The nexus between inflation and financial integration in the euro area. *Applied Economics Letters*, 32, 1–5. https://doi.org/10.1080/13504851.2024.2333457.

Le, T.N.-L., Abakah, E. J. A., and Tiwari, A. K. (2021). Time and frequency domain connectedness and spillover among fintech, green bonds and cryptocurrencies in the age of the fourth industrial revolution. *Technological Forecasting and Social Change*, 162, 120382. https://doi.org/10.1016/j.techfore.2020.120382.

Li, R., Li, S., Yuan, D., and Yu, K. (2020). Does economic policy uncertainty in the U.S. influence stock markets in China and India? time-frequency evidence. *Applied Economics*, 52(39), 4300–4316. https://doi.org/10.1080/00036846.2020.1734182.

Li, S., and Chen, Q. (2021). Do the Shanghai–Hong Kong & Shenzhen–Hong Kong stock connect programs enhance co-movement between the mainland Chinese, Hong Kong, and U.S. stock markets? *International Journal of Finance and Economics*, 26(2), 2871–2890. https://doi.org/10.1002/ijfe.1940.

Ling, D. C., Wang, C., and Zhou, T. (2020). A first look at the impact of COVID-19 on commercial real estate prices: asset-level evidence. *The Review of Asset Pricing Studies*, 10(4), 669–704. https://doi.org/10.1093/rapstu/raaa014.

Maghyereh, A., and Abdoh, H. (2020). Tail dependence between Bitcoin and financial assets: evidence from a quantile cross-spectral approach. *International Review of Financial Analysis*, 71, 101545. https://doi.org/10.1016/j.irfa.2020.101545.

Mensi, W., Sensoy, A., Aslan, A., and Kang, S. H. (2019). High-frequency asymmetric volatility connectedness between Bitcoin and major precious metals markets. *The North American Journal of Economics and Finance*, 50, 101031. https://doi.org/10.1016/j.najef.2019.101031.

Mensi, W., Vo, X. V., Ko, H. U., and Kang, S. H. (2023). Frequency spillovers between green bonds, global factors and stock market before and during COVID-19 crisis. *Economic Analysis and Policy*, 77, 558–580. https://doi.org/10.1016/j.eap.2022.12.010.

Monteiro, A. S., Sebastião, H., and Silva, N. (2024). Prediction and allocation of stocks, bonds, and REITs in the US market. *Computational Economics*, 6, 1–40. https://doi.org/10.1007/s10614-024-10589-2.

Nguyen, T. T. H., Naeem, M. A., Balli, F., Balli, H. O., and Vo, X. V. (2021). Time-frequency comovement among green bonds, stocks, commodities, clean energy, and conventional bonds. *Finance Research Letters*, 40, 101739. https://doi.org/10.1016/j.frl.2020.101739.

Okorie, D. I., and Lin, B. (2020). Did China's ICO ban alter the Bitcoin market?. International Review of Economics & Finance, 69, 977–993. https://doi.org/10.1016/j.iref.2020.05.016

Phillip, A., Chan, J. S. K., and Peiris, S. (2018). A new look at cryptocurrencies. *Economics Letters*, 163, 6–9. https://doi.org/10.1016/j.econlet.2017.11.020.

Reboredo, J. C., Ugolini, A., and Aiube, F. A. L. (2020). Network connectedness of green bonds and asset classes. *Energy Economics*, 86, 104629. https://doi.org/10.1016/j.eneco.2019.104629.

Saeed, T., Bouri, E., and Tran, D. K. (2020). Hedging strategies of green assets against dirty energy assets. *Energies*, 13(12), 3141. https://doi.org/10.3390/en13123141.

Sensoy, A. (2019). The inefficiency of Bitcoin revisited: a high-frequency analysis with alternative currencies. *Finance Research Letters*, 28, 68–73. https://doi.org/10.1016/j.frl.2018.04.002.

Seth, N., and Panda, L. (2018). Financial contagion: review of empirical literature. *Qualitative Research in Financial Markets*, 10(1), 15–70. https://doi.org/10.1108/QRFM-06-2017-0056.

Su, C.-W., Qin, M., Tao, R., and Zhang, X. (2020). Is the status of gold threatened by Bitcoin?. *Economic Research-Ekonomska Istraživanja*, 33(1), 420–437. https://doi.org/10.1080/1331677X.2020.1718524.

Suleman, M. T., Sheikh, U. A., Galariotis, E. C., and Roubaud, D. (2023). The impact of bitcoin fear and greed on good and bad network connectedness: the case of the U.S. sectoral high frequency returns. *Annals of Operations Research*, 338, 1–45. https://doi.org/10.1007/s10479-023-05455-7

Yang, L., Cai, X. J., Zhang, H., and Hamori, S. (2016). Interdependence of foreign exchange markets: A wavelet coherence analysis. Economic Modelling, 55, 6–14. https://doi.org/10.1016/j.econmod.2016.01.022.

Zhou, J., and Anderson, R. I. (2013). An empirical investigation of herding behavior in the US REIT market. The Journal of Real Estate Finance and Economics, 47, 83–108. https://doi.org/10.1007/s11146-011-9352-x

Zhu, B., and Lizieri, C. (2022) Local beta: has local real estate market risk been priced in REIT returns?. *The Journal of Real Estate Finance and Economics*, 69, 1–37. https://doi.org/10.1007/s11146-022-09890-4.

2 Investigation of green supply chain management practices in small scale industries: applying analytic hierarchy process method and paired correlation

PrashantKumar S. Bajaj[1,a], *Sanjay P. Shekhawat*[2,b] *and Pradeep M. Solanki*[3,c]

[1]Research Scholar, Mechanical Engineering, K.B.C. North Maharashtra University, Jalgaon, Maharashtra, India

[2]Professor, Mechanical Engineering, G H R Institute of Engineering and Business Management, Jalgaon, Maharashtra, India

[3]Assistant Professor, Mechanical Engineering, S.S.B.T. College of Engineering and Technology, Jalgaon, Maharashtra, India

Abstract

This paper deals with the green supply chain management (GSCM) and its impact on SSIs in the Jalgaon district, Maharashtra, India. It explores the GSCM practices, drivers, pressures, and barriers and their influence on the environmental, operational, and economic performance of SSIs. Survey questionnaires, statistical analyses, the analytic hierarchy process(AHP), paired co-relational studies, and reliability evaluations are all part of the process. The research findings show a significant correlation between GSCM Practices and better performance. Furthermore, GSCM drivers have a strong correlation with improved performance. The Analytic hierarchy process (AHP) analysis reveals the GSCM element prioritization. Among them, green practices, green design and green supplier emerge as top influencers. For GSCM drivers, industrial drivers and purchasing drivers claim the highest ranks. In the realm of GSCM barriers, technological barriers take the lead. Lastly, GSCM pressures regulatory pressures and customer pressures as top priorities. This study contributes to the understanding of GSCM's intricate relationships with the proposed GSCM model for sustainable business practices in small scale industries (SSIs). These practices will help the researcher to implement the GSCM to SSI.

Keywords: Analytic hierarchy process, consistency index, consistency ratio, green supply chain management, small scale industries

Introduction

Environmental management has become a topic of mutual concern for governments and consumers due to increasing high levels of industrialization (Bajaj, Bansod, and Paul, 2018). The growing concern in the global market for "green" issues and the scarcity of natural resources have forced executives to view supply chain strategies from an environmental perspective. High environmental risk industries, including chemical, plastic, automotive, and heavy engineering, have always considered improvements in environmental performance as one of the basic competitive priorities, alongside lower costs, manufacturing lead time, and quality. For instance, the European Union passed the Restriction of Hazardous Substance

[a]prashantsbajaj@gmail.com, [b]spshekhawat@gmail.com, [c]p_msolanki@rediffmail.com

DOI: 10.1201/9781003564447-2

and the Waste Electrical and Electronic Equipment regulations demanding compliance with the relevant regulatory laws of product recycling and prohibiting the use of hazardous substances in products for sale in the market. Thus this intends the SSI to implement GSCM in practices. Various authors tried to implement GSCM in SSI to give them a support to survive in this global world (Govindan, Azevedo, Carvalho, and Cruz-Machado, 2015, Hsu, et.al. 2013, Sarkis et.al. 2011). In India, small scale industries (SSIs) play an important role in contributing 40% to the economy and 45 to 50% to employment (Hariharan Ganeshan and P. Suresh, 2015). About 90% of SME and MSME have not registered with the State Pollution Control Board and Central Pollution Control Board. In this development, the supply chain manager plays an important role in selecting and developing appropriate green strategies to improve environmental, economic, and social performance as well as gain a competitive advantage.

Literature Review

Background of GSCM

During the early stages of the 20th-century industrial revolution, the concept of industrial pollution began to emerge. Henry Ford played a significant role in revolutionizing the automotive supply chain and organizational methods, introducing pioneering practices such as lean and JIT manufacturing. During this period, the primary objectives of JIT and supply chain management (SCM) were to enhance operational efficiency and reduce waste. It's important to note that waste reduction was primarily driven by economic considerations, aiming to minimize financial losses (Lai and Cheng, 2009). The concept of green supply chain management (GSCM) received greater attention during the 1980s. In the middle of 1990, investigations of SCM from an ecological point of view came into existence. Sarkis systematically integrated environmental considerations into purchasing, operations, marketing, logistics, and reverse logistics (Sarkis, 1995a, 1995b). Subsequently, more case studies and theoretical investigations were conducted to evaluate GSCM practices (Seuring and Müller, 2008). This historical progression demonstrates the evolving relationship between industry and the environment, paving the way for the integration of sustainable and environmentally conscious practices into SCM (Bajaj, Shekhawat, and Solanki, 2023).

Purpose of Study

The purpose of this study is to address the evolving environmental concerns that have emerged over recent decades, particularly the increasing demand for environmentally friendly products among consumers. SSI has an impact on the environment, prompting a global exploration for more sustainable industrial management systems. GSCM represents a relatively recent concept aimed at enhancing organizational capabilities within the realm of supply chain management. In the context of Indian SSI, environmental concerns have historically been treated with less seriousness, and knowledge regarding GSCM remains notably insufficient. The findings from this study will be instrumental in formulating a strategic roadmap for the sustainable growth of SSI in the Jalgaon District. GSCM offers common benefits towards sustainability, such as improved customer value, cost reduction, enhanced operational efficiency, and competitive advantage (Luthra, Garg and Haleem, 2013).

Importance of the Study

SSIs have significant effects on the environment. This forces nations around the world to hunt for new environmentally friendly industrial management systems. This gives the loop to change in SCM from start to end customer. GSCM is essential and needed for every industry in each region. In India, SSI plays an important role in contributing 40% to the economy and 45 to 50% to employment (Hariharan and Dr. P. Suresh, 2015). About 90% of SSIs are not registered with the Pollution Control Board. In this contest, the manager, shop floor in charge and supplier play an important role in developing appropriate green strategies to improve environmental, economic, and operational performance.

Research Gaps

Despite the growing importance of GSCM in SSIs (Srivastava, 2007), there is a paucity of research in emerging markets such as India. This research gap is acute in Jalgaon, Maharashtra, a region with a significant concentration of SSIs. Additionally, while the analytic hierarchy process(AHP) has been a widely utilized tool, its specific application to prioritize key GSCM elements in SSIs remains largely unexplored. Moreover, existing research on GSCM in SSIs has often overlooked the direct link between GSCM elements and their impact on key performance indicators. This omission highlights a crucial research gap in evaluating the performance implications of GSCM adoption in SSIs in Jalgaon, Maharashtra. Furthermore, the broader research landscape in GSCM has primarily focused on developed countries and specific industries. Given the unique economic and industrial landscape of Jalgaon, Maharashtra, there is a need to contextualize research findings from SSIs in this region and assess their applicability in other similar contexts.

Research Approach

The research approach includes research design undertaken for the data collection methods, questionnaire development, objectives and hypotheses. Furthermore, the AHP analytical framework, statistical tools utilized, and data analysis techniques are considered. The focus of this study is to determine the extent of implementation of GSCM practices in SSIs in the Jalgaon district. The study approach included both exploratory and descriptive research methodologies, resulting in a combination research design. To obtain the data, interviews are conducted with individuals including managers, supervisors, and suppliers.

Research Questions

RQ1. To what extent managers, shop floor in charge and suppliers of Indian SSI in JalgaonDistrict are aware of GSCM practices and their implementation?

RQ2. What are the key elements of GSCM practices, drivers, pressure and barriers in SSIs ofIndia?

RQ3. How can the Indian SSIs adopt GSCM Practices in their organizations to improve their environmental, economic and operational Performance?

Research Objectives

The following objectives are framed for the research work.

1. To study the current scenario of the GSCM in Jalgaon District India.
2. To identify the operational parameters of GSCM practices.
3. To develop a sector-specific and generalized green supply chain study.
4. To explore and assess the various parameters as drivers, barriers, and pressures that SSIs encounter in their efforts to incorporate and execute GSCM practices.

Sampling Methodology

To assess the extent of Green Practices implementation in the SSI of Jalgaon district, Maharashtra, India, a survey was carried out. Quantitative research was used to collect data from a large number of respondents through questionnaires. The proposed model was tested across eight different industry types, including automobile, chemical, food processing, electrical/electronics, plastic, cement, fabrication, and paper. Questionnaires were tailored for three distinct organizational levels: shop floor in-charge, manager or owner, and supplier. In this study, an offline survey was conducted to gather essential information. The survey involved the distribution of a questionnaire with a series of questions administered in person, face to face, in four different stages. Out of the 300 targeted companies, a total of 313 responses were received from 154 companies at three different levels. This sample size was deemed sufficient for drawing conclusions and conducting statistical analysis. The sampling was conducted using a convenient method, relying on individuals such as friends, relatives, consultants, Engineering college faculty members, entrepreneurs, and industry contacts to gather responses from various SSIs. In some cases, personal visits were made to meet the company owner or responsible person.

Methodology Used for Survey and AHP Analysis

In this study, a questionnaire was employed as a tool to assess GSCM. The questionnaire consisted of four sections. The first section focuses on gathering demographic information about the SSIs in the Jalgaon region. In the Section 2respondents were given literature related to GSCM and the benefits that can be obtained by implementing green practices. They were then asked to rate the factors considered for implementing green practices in their organization on a five-point scale. Section 3 was planned to find out the ranking of three broad elements of GSCM green drivers, green pressure and green barriers. This section has different questions for different levels as engineers/managers, shop floor I/C and suppliers. Section 4 precisely planned to discuss regarding the SSI's performance while implementing green practices. It has included the three-dimensional performance as environmental performance, economic performance and operational performance. An AHP method was then utilized to prioritize the driver, barrier, and pressure. Figure 2.1 shows categorization of GSCM practices, drivers, pressures and barriers mentioned with specific parameters, which are actually a part of questionnaires based on likert scale.

Hypothesis

H0-Null Hypothesis H1-Alternative Hypothesis

1. H10: There is no significant correlation between GSCM Practices & Performances.
   H11: There is a significant correlation between GSCM Practices & Performances.
2. H20: There is no significant correlation between GSCM Drivers & Performances.
   H21: There is a significant correlation between GSCM Drivers & Performances.
3. H30: There is no significant correlation between GSCM Barriers & Performances.
   H31: There is a significant correlation between GSCM Barriers & Performances.
4. H40: There is no significant correlation between GSCM Pressure & Performance.
   H41: There is a significant correlation between GSCM Pressure & Performance.

Results and Discussions

AHP Method for Analysis

Results of AHP framework analysis for understanding and implementation of GSCM principles, leading to sustainable and efficient supply chain practices.Table 2.1 presents the results of GSCM practices

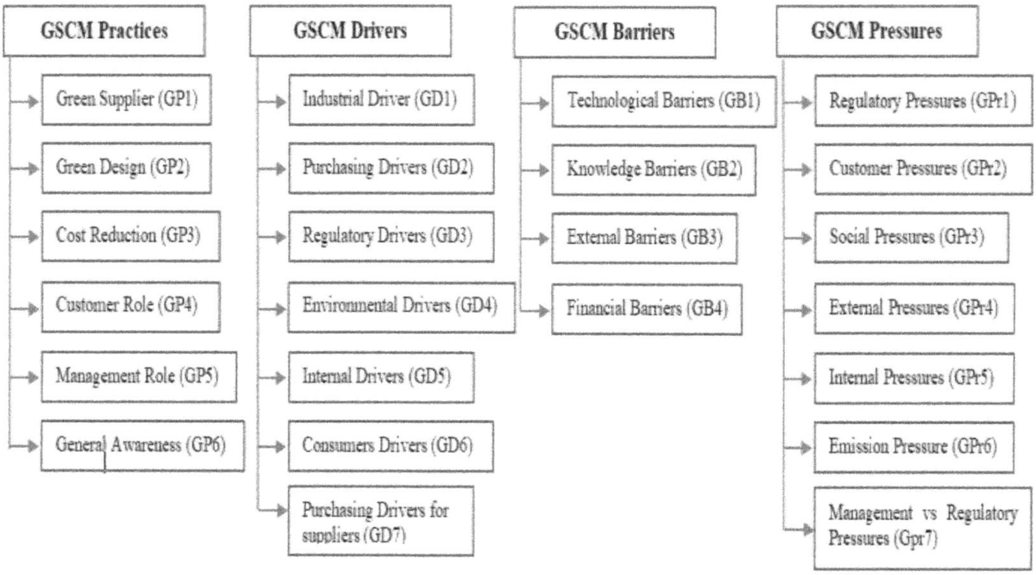

Figure 2.1 Categorization of Key GSCM Parameters
Source: Developed by Author

Table 2.1 Pair-wise matrix for GSCM practices.

Green practices	Green supplier (GP1)	Green design (GP2)	Cost reduction (GP3)	Customer role (GP4)	Management role (GP5)	General Awareness (GP6)	Normalized GM (Weights)
GP1	1.00	1.62	1.62	3.00	3.00	0.33	21.8%
GP2	0.62	1.00	1.62	3.78	2.00	1.74	23.7%
GP3	0.62	0.62	1.00	3.97	2.17	1.81	20.8%
GP4	0.33	0.14	0.25	1.00	0.57	0.57	6.15%
GP5	0.14	0.50	0.46	1.74	1.00	0.50	8.56%
GP6	3.03	0.57	0.55	1.74	2.00	1.00	18.9%

CI = 0.096, CR = 0.077, CR<0.1, RI = 1.25
Source: Developed by Authors

from GP1 to GP6. The consistency index (CI) is calculated as 0.096, which is less than 0.1, indicating an acceptable level of consistency.

The consistency ratio (CR) is calculated as 0.077, also falling below the threshold of 0.1. The random index (RI) is 1.25, providing a reference point for evaluating the consistency of the pairwise comparisons. Overall, the AHP analysis suggests that green design (GP2) holds the highest importance with a weight of 23.77%, followed by green supplier (GP1) with a weight of 21.80%. Other factors such as cost reduction (GP3), general awareness (GP6), management role (GP5), and customer role (GP4) also contribute to the GSCM practices, albeit with relatively lower weights

Table 2.2 presents the results of AHP analysis for different green drivers (GD1

Table 2.2 Pair-wise matrix for GSCM drivers.

Green drivers	Industrial drivers (GD1)	Purchasing drivers (GD2)	Regulatory drivers (GD3)	Environmental drivers (GD4)	Internal drivers (GD5)	Consumers drivers (GD6)	Supplier Awareness drivers (GD7)	Normalized GM (Weights)
GD1	1.00	6.00	6.00	5.00	4.00	3.00	3.00	39.18%
GD2	0.17	1.00	5.00	5.00	4.00	4.00	3.00	21.94%
GD3	0.17	0.20	1.00	1.00	0.50	0.50	0.50	4.91%
GD4	0.20	0.20	1.00	1.00	0.50	0.50	0.50	5.06%
GD5	0.14	0.25	2.00	2.00	1.00	2.00	0.50	8.84%
GD6	0.33	0.25	2.00	2.00	0.50	1.00	0.50	8.08%
GD7	0.33	0.33	2.00	2.00	2.00	2.00	1.00	12%

CI=0.066, CR=0.049, CR<0.1, RI = 1.35

Source: Developed by Authors

Table 2.3 Pair-wise matrix for GSCM barriers.

Green barriers	Technological barriers (GB1)	Knowledge Barriers (GB2)	External Barriers (GB3)	Financial barriers (GB4)	Normalized GM (Weights)
GB1	1.000	6.000	5.000	2.000	62.01%
GB2	0.167	1.000	3.000	0.333	14.23%
GB3	0.200	0.333	1.000	0.333	8.60%
GB4	0.500	0.143	3.000	1.000	15.16%

CI = 0.096, CR = 0.077 CR<0.1, RI = 1.25

Source: Developed by Authors

to GD7). Industrial drivers (GD1) hold the highest weight of 39.18%, followed by purchasing drivers (GD2) with a weight of 21.94%. Other factors such as internal drivers (GD5), supplier awareness drivers (GD7), consumers drivers (GD6), environmental drivers (GD4), and regulatory drivers (GD3) contribute to GSCM, albeit with relatively lower weights. The CI is calculated as 0.066, which is less than 0.1, indicating an acceptable level of consistency. The CR is determined as 0.049, falling below the threshold of 0.1. The RI is considered 1.35, providing a reference point for evaluating the consistency of the pairwise comparisons. Industrial drivers and purchasing drivers emerge as the most critical factors, followed by internal drivers, supplier awareness drivers, consumers drivers, environmental drivers, and regulatory drivers.

Table 2.3 shows AHP analysis for different GSCM barriers. Technological barriers (GB1) hold the highest weight of 62.01%, followed by knowledge barriers (GB2) with a weight of 14.23%. Other barriers such as financial barriers (GB4) and external barriers (GB3) contribute to GSCM but with relatively lower weights of 15.16% and 8.60% respectively. The CI is calculated as 0.096, indicating CR is determined as 0.077, falling below the threshold of 0.1. The RI is considered 1.25, providing a reference point for evaluating the consistency of the pairwise comparisons. The AHP analysis highlights the relative importance of different barriers within the GSCM framework. Technological barriers emerge as the most critical factor, followed by knowledge barriers, financial barriers, and external barriers.

Table 2.4 presents the results of the AHP analysis for different GSCM pressures. Regulatory pressures (GPr1) hold the highest weight of 30%, followed by customer pressures (GPr2) with a weight of 20%. Other pressures such as management to regulatory pressures (GPr7), social pressures (GPr3), external pressures (GPr4), emission pressure (GPr6), and internal pressures (GPr5) also contribute to GSCM, but with relatively lower weights. The CI is calculated as 0.046, which is less than 0.1, indicating an acceptable level of consistency. The CR is determined as 0.034, falling below the threshold of 0.1. The RI is considered 1.35, providing a reference point for evaluating the consistency of the pairwise comparisons. Regulatory pressures and customer pressures emerge as the most critical factors, followed by management regulatory pressures, social pressures, external pressures, emission pressures, and internal pressures.

Thus, the analysis provides insights for decision-makers to understand the key practices, drivers, barriers and pressures that drive the adoption of sustainable and environmentally responsible practices within the supply chain. By prioritizing their efforts organizations can effectively

Table 2.4 Pair-wise matrix for GSCM pressure.

Pressures (Weights)	Regulatory pressures (GPr1)	Customer pressures (GPr2)	Social pressures (GPr3)	External pressures (GPr4)	Internal pressures (GPr5)	Emission pressure (GPr6)	Management regulatory pressures (Gpr7)	Normalized GM
GPr1	1.00	4.00	3.00	3.00	2.00	3.00	3.00	30%
GPr2	0.25	1.00	3.00	3.00	4.00	2.00	0.33	20%
GPr3	0.33	0.33	1.00	1.00	2.00	0.50	0.50	8%
GPr4	0.33	0.33	1.00	1.00	1.00	2.00	0.33	9%
GPr5	0.14	0.25	0.50	1.00	1.00	3.00	2.00	7%
GPr6	0.33	0.50	2.00	0.50	0.33	1.00	0.33	8%
Gpr7	0.33	3.00	2	3	0.5	3	1	18%

CI = 0.046, CR = 0.034, CR<0.1, RI = 1.35

Source: Developed by Authors

Table 2.5 Reliability statistics.

	Cronbach's alpha	Number of items
GSCM practices	0.894	GP1,GP2,GP3,GP,GP5,GP6
GSCM drivers	0.846	GD1,GD2,GD3,GD4,GD5,GD6,GD7
GSCM barriers	0.816	GB1,GB2,GB3,GB,GB5
GSCM pressures	0.810	GPr1, GPr2, GPr3, GPr4, GPr5, GPr6, GPr7
GSM performance	0.872	GPe1,GPe2,GPe

Source: Developed by Authors

respond to external and internal demands and improve their overall environmental performance.

Selection of Factors Based on AHP Analysis
In this study, four factors from each table are selected based on their weights as follows:

GP2 - 23.77%, GP1 - 21.80%, GP3 - 20.81%, GP6- 18.92%
GD1 - 39.18%, GD2 - 21.94%, GD7 - 12.00%,GD - 8.84%,
GB1- 62.01%, GB4- 15.16%, GB2 - 14.23%, GB3 - 8.60%
GPr1- 30.00%, GPr2- 20.00%, GPr7- 18.00%, GPr4- 9.00%

Table 2.6 Correlation of GSCM elements with the performances.

GSCM elements	GPe1	GPe2	GPe3	Remark
GSCM practices				
GP1	-0.034	-0.027	-0.055	No strong correlation
GP2	-0.042	-0.019	-0.076	
GP3	0.125	0.081	0.013	
GP4	-0.012	0.025	-0.018	
GP5	0.119	0.072	0.02	
GP6	0.024	0.012	-0.019	
GSCM drivers				
GD1	-0.114	-0.091	0.033	No strong correlation
GD2	0.121	0.075	0.035	
GD3	0.021	-0.022	0.006	
GD4	-0.183	-0.192	-0.136	
GD5	-0.038	-0.051	0.007	
GD6	-0.005	-0.063	0.09	
GD7	-0.1	-0.096	-0.054	
GSCM barrier				
GB1	-0.065	-0.081	-0.102	No strong correlation
GB2	-0.124	-0.153	-0.104	
GB3	-0.135	-0.102	-0.085	
GB4	-0.09	-0.115	-0.117	
GB5	-0.13	-0.132	-0.074	
GSCM pressures				
GPr1	-0.009	-0.031	0.076	No strong correlation
GPr2	0.085	0.089	0.139	
GPr3	-0.06	-0.093	0.107	
GPr4	0.095	0.024	0.106	
GPr5	0.083	-0.006	0.057	
GPr6	-0.086	-0.061	0.009	
GPr7	-0.101	-0.101	0.069	

Source: Developed by Authors

Reliability Statistics

The analysis of Cronbach's Alpha coefficients offers valuable insights into the reliability of the measurement scales employed in the study. Higher coefficients not only denote enhanced internal coherence but also underscore the scales' ability to accurately capture the underlying constructs. This enhances the credibility of the study's measurements and reinforces the validity of the interpretations drawn from these measurements as mentioned in Table 2.5. For GSCM practices (GP), the robust Cronbach's alpha coefficient of 0.894 highlights a strong internal harmony among the 6 items within the scale. Similarly, the coefficients of 0.846 for GSCM Drivers (GD), 0.810 for GSCM Pressures (GPr), and 0.816 for GSCM Barriers (GB) also exhibit decent internal consistency.

Correlation Pair Wise Matrix:

Table 2.6 presents the correlation analysis of various GSCM elements with different performance indicators. The Table displays correlation coefficients between GSCM Practices (GP), GSCM Drivers (GD), GSCM Barriers (GB), GSCM Pressures (GPr), and performances (GPe1, GPe2, GPe3). Here is a summary of the findings:

No strong correlation between practices and drivers, barriers, and pressures has been established as noted by the correlation coefficients precise to 0. However, drivers show a weak positive correlation with GPe3 with a correlation coefficient of 0.003. It has a weak positive correlation with the total, with a correlation coefficient of 0.148. This indicates a minor positive tendency for GSCM drivers to be associated with overall performance improvement.

Conclusion

The analytic hierarchy process(AHP) analysis has provided valuable insights into the critical aspects of green supply chain management (GSCM). These findings offer a comprehensive framework to enhance their environmental sustainability practices. For specific GSCM practices, factors such as green design, green suppliers, cost reduction, customer roles, and management roles have been identified as pivotal. Regarding GSCM drivers, regulatory, customer, governmental, purchasing, environmental, internal, and consumer drivers have been recognized as key motivators for GSCM adoption. In terms of GSCM barriers, technological, knowledge, external, and financial barriers have been identified as significant challenges. Lastly, GSCM pressures, including regulatory, customer, social, external, internal, emission management, and management vs. regulatory pressures, highlight the multifaceted motivations behind environmental practices. Incorporating these recommendations and insights will enable organizations to effectively implement GSCM practices, overcome barriers, and navigate the complex landscape of pressures, ultimately contributing to a more sustainable and environmentally responsible future.

Implications and Recommendations

The findings of this analysis have significant implications for Managers seeking to implement effective GSCMpractices. Some specific recommendations for organizations include:

- Partner with green suppliers and implement green procurement practices.
- Design eco-friendly products and services by considering environmental factors throughout the product development process.
- Implement waste reduction, recycling, and efficient equipment utilization strategies to reduce environmental impact and optimize costs.
- Educate customers about green products and services, and communicate the

organization's commitment to environmental sustainability.
- Foster a culture of environmental stewardship within the organization by empowering top management to lead the greening initiatives, making quick decisions at all levels, training employees on green practices, obtaining ISO 14001 certification, conducting regular audits, and monitoring environmental performance.

The analysis of GSCM practices reveals a wide range of practices that can be adopted to enhance the environment. These practices can be broadly categorized into green supplier, green design, cost reduction, customer role, and management role practices. By taking these recommendations into account, organizations can effectively implement specific GSCM practices to achieve their environmental sustainability goals and build a more sustainable future for all.

References

Bajaj, P. S., Bansod, S. V., and Paul, I. D. (2018). A review on the green supply chain management (GSCM) practices, implementation and study of different framework to get the area of research in GSCM. In Techno-Societal 2016: Proceedings of the International Conference on Advanced Technologies for Societal Applications, (pp. 193–199). Springer International Publishing.

Bajaj, P. S., Shekhawat, S. P., and Solanki, P. M. (2023).Practices and drivers for green supply chain management implementation among small scale industries. Industrial Engineering Journal, 52(10), 3.

Chen, Y.-S. (2008). The driver of green innovation and green image–green core competence. Journal of Business Ethics, 81, 531–543.

Govindan, K., Azevedo, S. G.,Carvalho, H., and Cruz-Machado, V.(2015). Lean, green and resilient practices influence on supply chain performance: an interpretive structural modellingapproach. International Journal of Environmental Science and Technology, 12, 15–34.

Hariharan Ganeshan and P.Suresh.(2015). Implementation of green supply chain in small scale industries.International Journal of Applied Engineering Research, 10(19), 14903–14915.

Hsu, C. C.,Choon Tan, K., HanimMohamadZailani, S., and Jayaraman, V. (2013). Supply chain drivers that foster the development of green initiatives in an emerging economy. International Journal of Operations and Production Management, 33(6), 656–688.

K. H. Lai., Yee, Rachel WY, Andy CL Yeung, and TC Edwin Cheng, (2009) "The service-profit chain: A review and extension." Total Quality Management 20, no. 6: 617-632.

Luthra, S., Garg, D., and Haleem, A. (2013). Identifying and ranking of strategies to implement green supply chain management in Indian manufacturing industry using analytical hierarchy process. Journal of Industrial Engineering and Management, 6(4), 930–962.

Sarkis, Joseph. (1995) "Manufacturing strategy and environmental consciousness." Technovation 15, no. 2: 79-97.

Sarkis, Joseph, and Abdul Rasheed. (1995) "Greening the manufacturing function." Business Horizons-Bloomington- 38: 17–17.

Sarkis, J., Zhu, Q., and Lai, K.(2011).An organizational theoretic review of green supply chain management literature.International Journal of Production Economics, 130(1), 1–15.

Srivastava, S. K.(2007). Green supply chain management: a state of the art literature review. International Journal of Management Reviews, 9(1), 53–80.

Searing, M. J., & Müller, M. (2008). A framework for evaluating green supply chain management performance. Journal of Cleaner Production, 16(15), 1565–1579.

3 The impact of eco-brand expansion strategy on brand engagement concerning the fashion industry: a bibliometric review

Truptimayee Das[a], *Akankshya Patnaik*[b] *and Samir Kumar Mapatro*[c]

Department of Management, NIST University, Berhampur, Odisha, India

Abstract

The growing importance of using environmentally friendly branding in the fashion industry highlights a significant shift towards making environmental sustainability a central part of managing fashion brands. However, existing research has different views on these green strategies' effectiveness. In this chapter, we use the resonance theory to suggest ideas to improve the relationship between consumers, brands, and environmental sustainability in eco-friendly fashion branding. We explore how (1) consumer behavior aligns with being environmentally friendly, (2) brand dynamics interact with environmental sustainability, and (3) the interactions between consumers and brands affect the acceptance and decision-making around green fashion brands. We do this by looking at bibliometric analysis, which helps us understand the critical connections. The degree to which customers comprehend and assimilate new environmental information as part of the company's broader strategy determines the success of eco-friendly brand activities. People's feelings about environmentally conscious brands are influenced by how well the brand incorporates environmental sustainability into its overall identity. Factors like how well the brand aligns with ecological values and how genuine the company's commitment to sustainability also play a significant role in how consumers respond. Moreover, how much consumers embrace the green attributes of a brand depends on their capacity to connect the fashion brand's name with eco principles and their involvement with the company. In this review paper, the author explored the impact of green branding on the fashion industry and suggested directions for future research.

Keywords: Eco-fashion, eco-friendly brand, engaging with the brand, environmental sustainability, expanding brand

Introduction

There is a growing trend in marketing to consider environmental issues. In recent years, businesses have had to rethink their belief that the environment will always provide resources without consequences and that their actions will not affect the environment (Olsen et al., 2014). Customers increasingly seek eco-friendly or "green" items to support their sustainable lives due to environmental signs, particularly concerning global warming. To mitigate the effects of production, design, packaging, and consumption on the environment, the marketing sector is shifting toward a greener strategy (Delafrooz et al., 2014). Sustainability has become a priority for organizations to ensure steady business expansion. Businesses focusing on sustainability initiatives have improved financial performance and consumer relations. Sustainable business strategies can also cut operating expenses and give businesses a

[a]truptimayeedas1990@gamil.com, [b]akankshya@nist.edu, [c]mahapatro.samir1991@gmail.com

DOI: 10.1201/9781003564447-3

competitive edge. Sustainability involves an organization's economic, social, and environmental responsibilities, raising the question of how to implement them. Companies must balance these three obligations, considering immediate and long-term effects. They should also consider reducing their environmental impact, providing opportunities for their staff, and supporting local communities (Ki and Kim, 2016). Olsen et al. (2014) analyzed the lasting effects of releasing new green products on brand attitude in the quickly evolving consumer goods industry and discovered a favorable association. Numerous businesses that use brand extension tactics to enter the green product market and already have a well-known brand have also looked at what influences how green brand extensions are evaluated. While Chatterjee (2009) found that the analysis of eco-brand extensions was more for products with a higher environmental impact. According to studies conducted, green brand extensions are viewed more favorably when there is a perceived fit between the brand and the extension as well as between the brand and the environmental cause that is being supported. According to Sang (Jain and Mishra, 2020), sustainable practices in the traditional fashion market improve customer satisfaction, brand perception, and trust while favorably influencing brand loyalty growth.

Brand extension studies, generally well-established in marketing research center on researcher success criteria when entering a new product category or market with an established brand name (Lee et al., 2020; Ki and Kim, 2016). Customers may still choose their favorite brand above environmentally friendly products, but their purchase habits may change as they become more conscious of sustainability (Putri and Yeshika, 2024). Businesses wishing to prosper over the long term must employ green branding techniques such as GBEs. Given the growing interest in brand extension strategies and the increased attention to environmental issues, it is surprising that the factors impacting assessments of brand extensions, mainly green brand extensions, have yet to be widely investigated in diverse contexts. Consequently, it is imperative to examine these characteristics to gain a more profound understanding of how brand extensions affect consumer loyalty and engagement. Olsen et al. (2014) Many studies have been conducted on how brand extensions affect consumers' perceptions of a brand. However, more is needed to know about the impact of newly introduced environmentally friendly products on customer attitudes toward such brands. Future research should closely examine customers' pro-environment views and how they affect brand loyalty (Di Leo et al., 2023). More studies need to be done on the connection between brand loyalty and a brand's green image (Chatterjee, 2009).

Concerning effects of rapid economic expansion include resource depletion, environmental degradation, and worsening climate change. As a result, companies are under increasing pressure to shift their manufacturing and sales processes for clothing toward ecologically friendly ones. It is now essential to use behavioral methods to encourage changes in consumption patterns toward environmentally friendly decisions (Delafrooz et al., 2014). By adopting these tactics, businesses can reduce anthropogenic environmental constraints and achieve green competitive advantages in home and international markets. To provide meaningful competitive advantages for businesses, the article will analyze trends in research development throughout different periods. As far as our knowledge is concerned, no such study has used the bibliometric method to study the topic we have taken. By developing new research directions, this study seeks to improve earlier research, close gaps in the literature, and broaden our understanding of eco-fashion. Its specific objectives were to define an investigation agenda, identify possible future research lines, and examine the global advancement of "eco-brand expansion strategy" research as published in specialized journals. The study will focus on answering the following research questions.

RQ1: How do fashion firms' green brand extensions impact customers' engagement and loyalty to their parent brands?
RQ2: What patterns, themes, and crucial information surface from a thorough bibliometric investigation of the fashion industry's green brand extension, engagement, and loyalty?
RQ3: What are the most common keywords, and how can cluster and co-word analyses be done?

Literature Review

This comprehensive literature analysis investigates theories concerning brand engagement and loyalty and the importance of green brand extension in the fashion sector. The chapter starts by defining "green brand extension" and then looks at how it affects marketing and performance in the fashion industry. It also assesses different brand extensions and explores the literature on green branding. According to Kapoor et al. (2023), brand extensions introduce new items under a well-known brand name within a specific product category, comprising novelties such as new tastes, forms, or container sizes. This approach, which makes up most brand extensions, fulfills many objectives, such as satisfying consumer needs for diversity and promoting sustainability. Fashion, defined by high impulse buying levels, shorter product life cycles, and erratic market demand, is vital in forming societal perceptions and interactions (Putri and Yeshika, 2024). As a result, companies—fast fashion in particular—are increasingly embracing sustainable branding techniques to appeal to customers who care about the environment and society (Henninger et al., 2016). To address repressed customer demands for sustainability, green brand extension initiatives use well-known brands to present eco-friendly substitutes (Chatterjee, 2009). According to studies, these brand extensions give businesses a low-risk and efficient way to break into new markets while enhancing consumer demand for environmentally friendly goods (Ki and Kim, 2016). According to Lee et al. (2020), there are generational differences in how baby boomers, generation X, generation Y, and generation Z view sustainability in the fashion sector. Knowing how customers behave is crucial, particularly in omnichannel purchasing, when judgments about what to buy are influenced by perceived value, repurchase intentions, and satisfaction recognition (Sinha et al., 2023). Although ethical and environmental factors have historically played a minor role in fashion purchases, current trends favor sustainability (Kim et al., 2022). Younger generations are more aware of environmental issues, which influence their clothing choices (Marcella-Hood, 2023).

This study aims to clarify four essential components of brand expansion: functional attributes, brand understanding, brand fit, and emotional connection. Emotions play a significant role in determining how consumers feel about brands and more positive attitudes toward brand extensions are linked to better brand effects (Di Leo et al., 2023). Environmental happiness and other elements, including practical, sentimental, and altruistic motives, impact brand loyalty, particularly regarding sustainable products (Retamal et al., 2023). Fashion brands may meet the increasing demand for sustainability from environmentally conscious customers by using green brand extension strategies that highlight brand fit, emotional connection, knowledge, and functional attributes as crucial factors in shaping consumer attitudes and loyalty. Understanding these dynamics is essential for firms to navigate the junction of sustainability and consumer desires as the fashion industry changes. Higher levels of brand effect are associated with more positive attitudes toward brand extensions, emphasizing the importance of emotions in assessing brand extensions (Kapoor et al., 2023). Brand love, the degree of emotional attachment a satisfied consumer has for a particular brand, positively impacts affective brand commitment and can influence the desire for impulsive purchases (Delafrooz et al., 2014). The effectiveness of a brand extension is often determined by the fit between the parent brand and its extension. As explained by Geegamage et al. (2023), brand knowledge encompasses two dimensions: brand awareness and brand image. Brand awareness involves customers' ability to recognize a specific brand in various situations, while brand image reflects perceptions about a brand held in consumers' memory (Ki and Kim, 2016). Shop or product attributes shape a brand's perception, and sustainability initiatives by fashion brands significantly impact brand image (Olsen et al., 2014). Consumers form a favorable brand attitude when they perceive a brand to have high functionality, indicating satisfactory quality (Wang et al., 2019). In the fashion sector, consumers consider style, fit, color, and fabrication as functional brand qualities (Yao et al., 2020).

Brand loyalty is a deeply held commitment to consistently repurchase or patronize a preferred product or service despite situational influences and marketing efforts that may lead to switching behavior (Marcella-Hood, 2023). Regarding sustainable products, customers' environmental satisfaction is correlated with the pleasure they derive from using them when fulfilling their needs and expectations (Retamal et al., 2023). Functional, emotional, and altruistic reasons influence brand loyalty to environmentally friendly products, particularly in the context of GBEs, as they encourage customers to pay more for green products (Jain and Mishra, 2020). The brand effect correlates with more favorable attitudes toward brand extensions, highlighting the role of emotions in assessing them. Brand love, reflecting consumers' emotional attachment to a brand, positively influences affective brand commitment and impulsive buying tendencies. The fit between a parent brand and its extension often determines the success of the extension. Brand knowledge encompasses brand awareness and image, influencing consumer perceptions (Geegamage et al., 2023). Attributes like sustainability significantly impact brand image, shaping consumer attitudes. In fashion, functionality, including style and fit, drives favorable brand attitudes. Brand loyalty entails consistent patronage despite external influences (Sinha et al., 2023). For sustainable products, environmental satisfaction correlates with consumer pleasure (Hill and Lee, 2015). Green brand extensions cater to environmentally conscious consumers, fostering loyalty through functional, emotional, and altruistic appeals. Understanding emotional connections, brand fit, knowledge, and functional qualities is crucial for fashion brands to navigate sustainability and consumer preferences (Wang et al., 2019). The study emphasizes the significance of emotional connection, brand fit, brand knowledge, and functional attributes in influencing consumers' attitudes and loyalty toward environmentally friendly brand extensions. As the fashion industry evolves, understanding these dynamics becomes imperative for brands seeking to navigate the intersection of sustainability and consumer preferences. Using green brand extension strategies allows fashion brands to tap into the growing demand for sustainable choices among environmentally conscious consumers.

Methodology

This article employs a bibliometric analysis to answer the questions. In addressing the

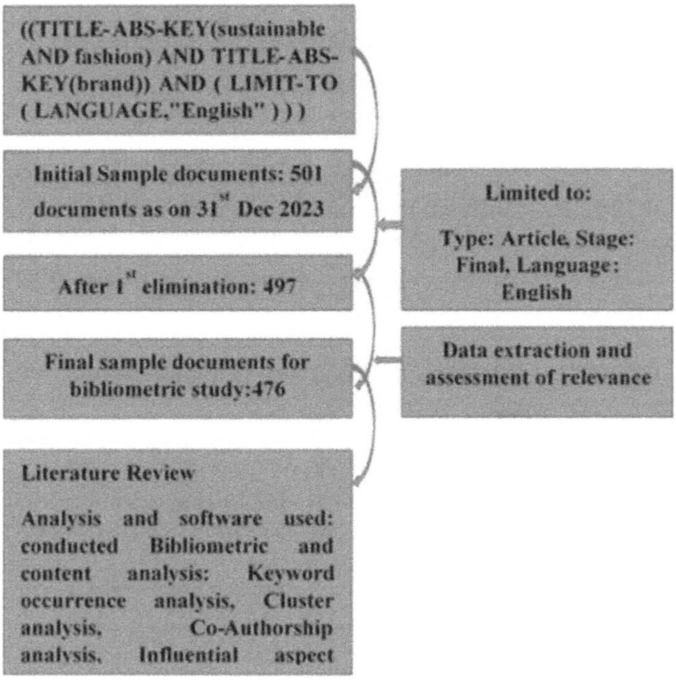

Figure 3.1 Research methodology
Source: Authors Compailation

research question, careful consideration was given to selecting the documents for review and determining the data extraction and analysis methodology as mentioned in the Figure 3.1. The decision to utilize the Scopus database stemmed from its recognition as the largest repository of peer-reviewed literature, particularly in business and management, surpassing the Web of Science (WOS) coverage (Chadegani et al., 2013). This choice was made to minimize the risk of overlooking essential documents for the Study. Subsequently, bibliometric information for all sample documents was downloaded in CSV format from Scopus, focusing on items released on or before December 31, 2023. We used the following keywords to get the data ([Title-ABS-key [sustainable and fashion] and TITLE-ABS-key [brand] and [limit-to language], English"). We collected around 497 articles, applying the inclusion and exclusion criteria of articles published in English and excluding other documents and articles published in different languages. Finally, we accepted 476 articles for the bibliometric analysis. The VOS viewer program was employed for network analysis, utilizing the CSV file as the input.

Additionally, an XLS file derived from the CSV data was imported to validate the relevance of each article and conduct supplementary bibliometric analysis. The authors harnessed widely recognized bibliometric indices, such as publication and citation counts, for their analysis. Figures representing the results were visually crafted using VOS viewer and MS Excel.

Analysis

Top Ten Publishing Countries

Figure 3.2 displays the top ten countries with the highest productivity. The leading country has an impressive output of over

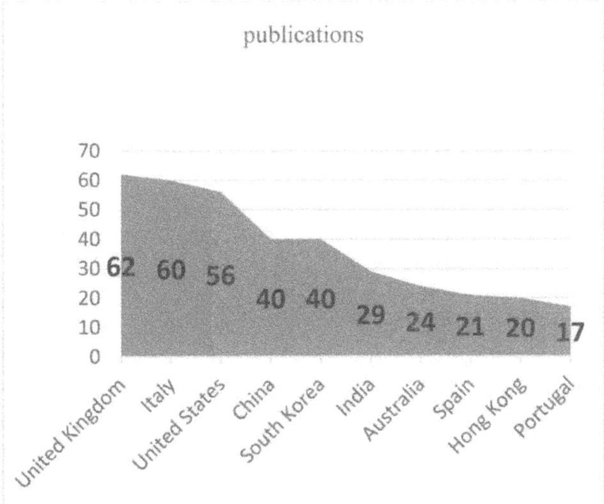

Figure 3.2 Top ten publishing countries
Source: Authors Compailation

369 publications, while the second and third countries have more than 122 publications each. The cumulative publications for the top five countries on the list exceed 258 during the study period. Notably, the UK stands out with remarkable productivity, contributing 62 publications, closely followed by Italy with 60 publications. The USA, China, Korea, India, and other notable countries follow suit in the rankings.

Most Eminent Authors

Figure 3.3 highlights the most eminent authors in this subject area. Henninger, C.E., Ko, E., and Nayak, R. stand out as the top three authors, each having published seven articles. Following closely is Arrigo, E., with six articles, and Blazquez, M., with four articles. The graph also depicts other distinguished authors and their corresponding number of publications.

Most Effective Journals

The diagram shows the top ten journals that were considered for analysis. The "Journal of Sustainability Switzerland" leads the field with 55 pertinent articles. "Environmental Footprints" and "Eco Design of Products and Processes" come in second with 19 papers. There is room for more research because other journals, such as the "Journal of Fashion Marketing and Management", contribute substantially. Focusing on these journals can increase the effect and reach of publications in the future. Figure 3.4.

Table 3.1 lists several academic articles, their authors, and the number of citations each article has received. In academia, a highly cited author or article refers to a researcher or a piece of scholarly work garnered substantial citations from other academic publications. Citations indicate how different researchers have referenced or used the work in their studies, implying its influence, importance, and impact within the educational community.

Overlay Visualization of Co-Author Affiliation

This section illustrates using VOS viewer software to analyze co-authorship networks, as Van Eck and Waltman (year) outlined. Of the 427 authors in the pool, only 11 satisfied the requirements for collaboration, so a network map had to be made. Although the ties between the authors were

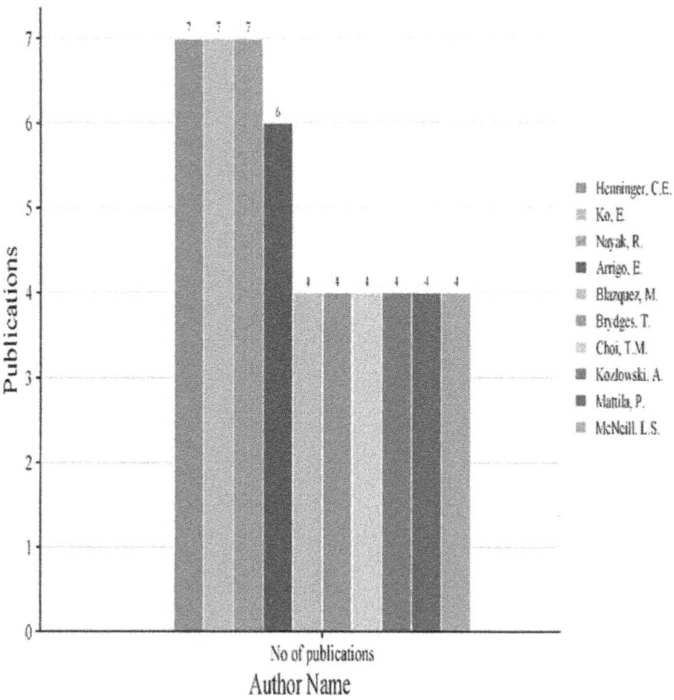

Figure 3.3 Most eminent authors
Source: Authors Compailation

Table 3.1 Top five highly cited articles with authors.

Sl no	Author with year	Article title	No of citation
1	Caniato et al., (2012)	Environmental sustainability in fashion supply chains: exploratory case-based research	368
2	Nagurney and Yu (2012)	Sustainable fashion supply chain management under oligopolistic competition and brand differentiation	178
3	Athwal et al. (2019)	Sustainable luxury marketing: a synthesis and research agenda	125
4	Han J et al. (2017)	Staging luxury experiences for understanding sustainable fashion consumption: a balance theory application	111
5	Bertola and Teunissen (2018)	Fashion 4.0. innovating the fashion industry through digital transformation	109

Source: Authors Compailation

not immediately apparent, this emphasizes how vital author collaboration is to advance research and creating influential papers. These cooperative initiatives promote knowledge sharing, multidisciplinary viewpoints, and the spread of creative ideas, eventually aiding in advancing academic pursuits. Figure 3.5.

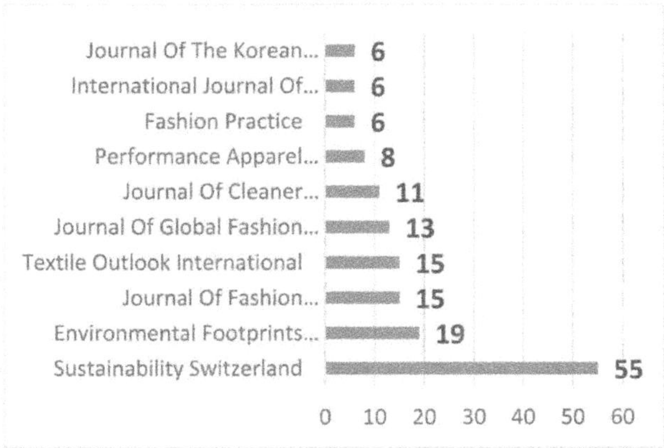

Figure 3.4 Most effective journals
Source: Authors Compailation

Co-Occurrence of Author Keyword Visualization

This part of the study can help readers understand how authors used specific words. They analyzed the amount of text and the quality of their work. One of the researchers, Zou (2019), aimed to group the words used in this research and other studies. This grouping helped researchers see what topics other studies found interesting. They looked at how words were connected and used together in articles. One can use VOS viewer software to analyze the keywords and their importance in studies by analyzing the frequency co-occurrence of words. Another research study By Zeng et al. (2010) used a similar analysis to show the connections between words used for grouping. This analysis helped put words into different groups. The size of the shapes used in their Study showed certain words' importance. Similar colors in the figure words belong to the same group. They used specific methods and tools to analyze the data and found 48 essential keywords out of 1318. They created seven groups related to Eco brand expansion strategy and brand value.

Each group of keywords showed a different aspect of eco branding strategy and its effects on brand value. For example, one group talked a lot about "Innovation" and "Sustainable Consumption," another about "Sustainable Luxury" and "Brand Image," and another about "Sustainability." Moreover, in "Fashion," each group had strongly connected words. For instance, "Sustainability" appeared the most frequently in one group. Another group used "Sustainable fashion" most often. This showed how researchers often used specific words in their studies about eco-brand expansion strategy and brand value. Figure 3.6.

Co-word Analysis

In our study of 476 research articles focusing on eco-brand expansion and brand engagement, the authors' 48 and 51 indexed keywords were identified by applying the criteria of minimum occurrence of 10 keywords. The primary author keywords, sustainability, fashion, and sustainable fashion, emerged as the top three regarding co-occurrences, as detailed in the figure below. Conversely, the top three indexed keywords from the research articles were Sustainable Development, Sustainability, and Textile, which occurred maximum time in the index keyword network map depicted in Figure above. An analysis of these keywords from both sources suggests that the authors emphasized sustainability

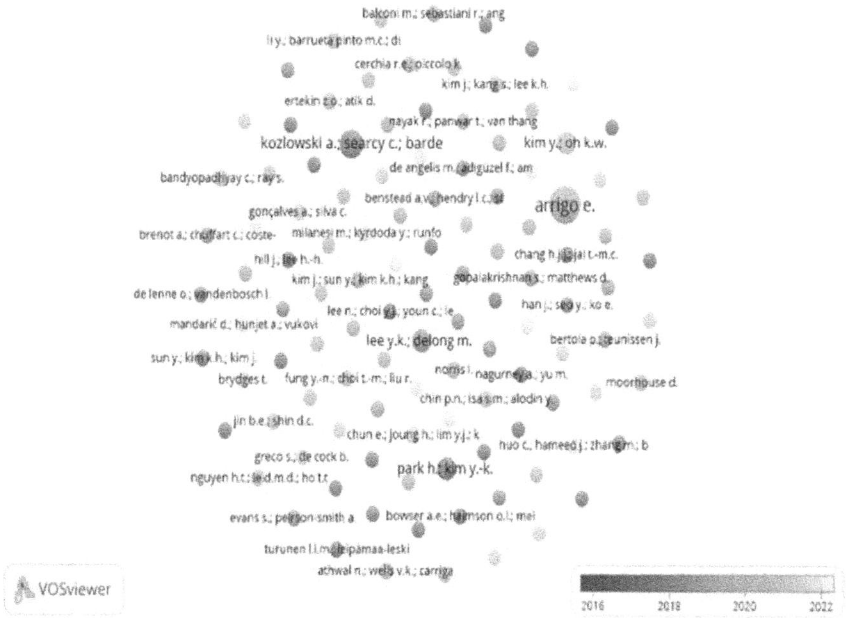

Figure 3.5 Network map of author and co-author analysis
Source: Authors Compailation

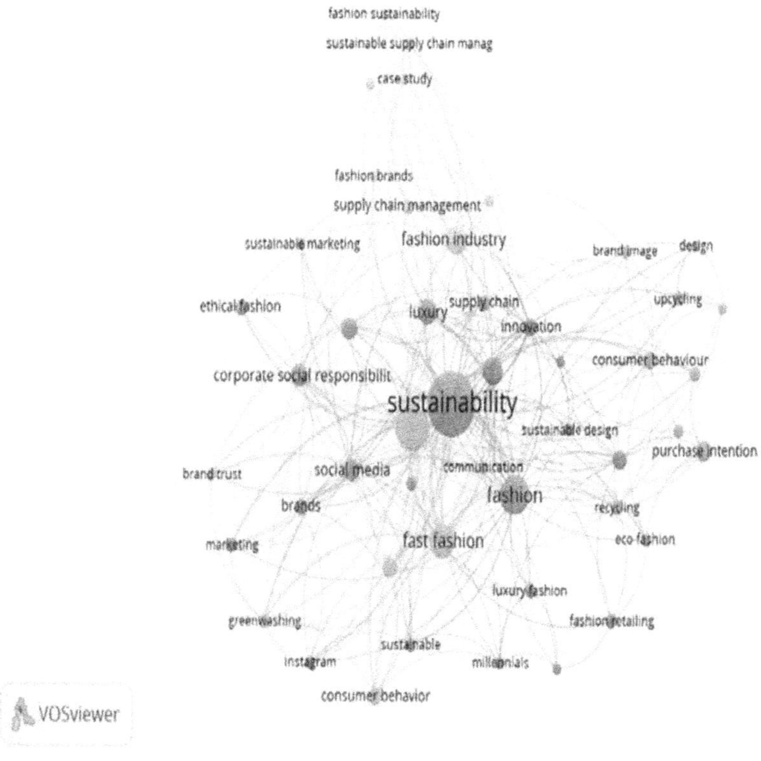

Figure 3.6 Author keyword co-occurrence network-map
Source: Authors Compailation

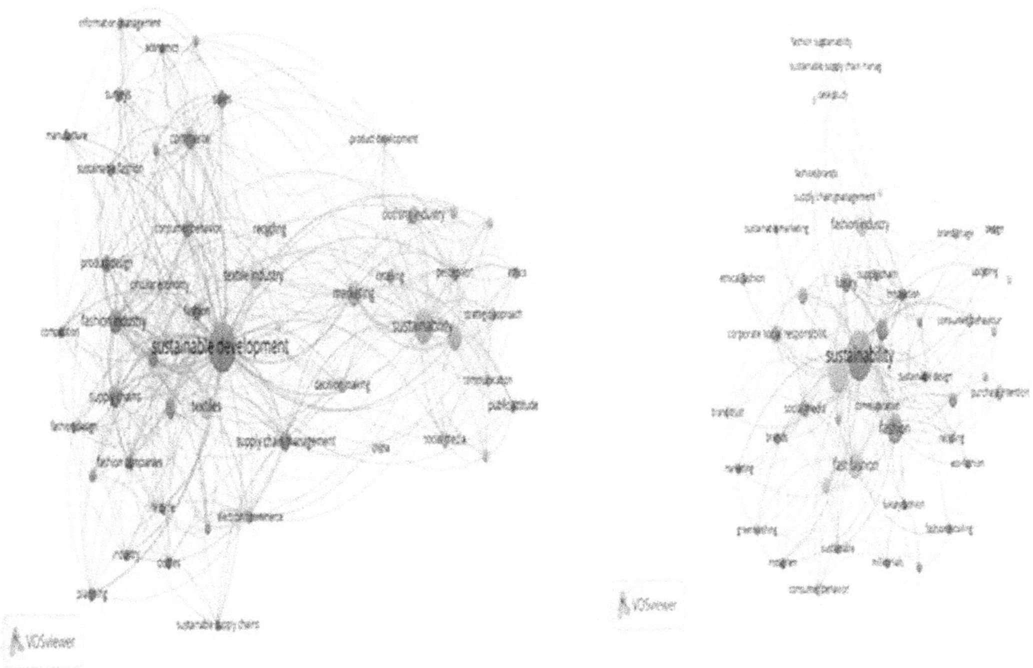

Figure 3.7 Co-word analysis
Source: Authors Compailation

in the fashion industry and its impact on developing brand image and the organization's profitability. Conversely, indexers focused more on sustainable development in the textile industry and, therefore, tried to ignite the development in social, economic, and organizational betterment. Figure 3.7.

CSR emerges as a prominent theme, with studies examining its impact on brand loyalty and overall engagement. The analysis also identifies a growing emphasis on consumer behavior within sustainable fashion, illustrating a shift in focus from traditional branding approaches to more environmentally conscious and socially responsible strategies.

Cluster Analysis

As it is clear from Figure 3.8 and Table 3.2, Cluster 1 is concerned with how design methods and customer behavior interact with sustainability in mind. It covers innovation, the circular economy, and ecologically friendly fashion design. In the sustainability framework, Cluster 2 investigates the relationship between luxury fashion brands and customer purchasing intentions. It contains terms associated with customer behavior toward luxury and eco-fashion brands, brand image, and communication. The third cluster focuses on the fashion industry's application of sustainability concepts. It addresses topics like social media interaction, marketing tactics, and the rise of sustainable fashion labels. The fourth cluster looks at the fashion sector's management techniques and supply chain procedures, focusing on sustainability issues. It has CSR, transparency, and sustainable supply chain management-related terms. Cluster 5 is devoted to corporate social responsibility (CSR) programs and how they affect sustainable development in the fashion industry. It covers subjects including sustainable business models, ethical fashion, and marketing strategies

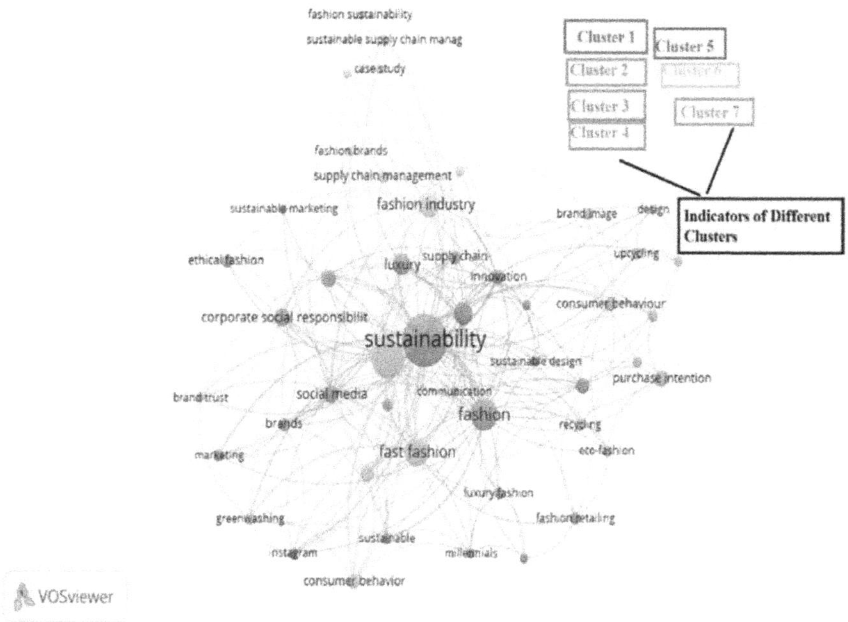

Figure 3.8 Cluster map
Source: Authors Compailation

Table 3.2 Cluster classification.

Cluster number	Cluster name
1	Sustainable Design and Sustainable Consumption
2	Purchase Intension and Luxury Brands
3	Fashion and Sustainability
4	Supply Chain Management and Fashion Industry
5	CSR and Sustainable Development
6	Sustainable Fashion and Consumer Behaviour
7	Sustainable Clothing and Design

Source: Authors Compailation

that support environmental objectives. The sixth installment explores customer views and behavior patterns around sustainable fashion techniques. It addresses issues like the use of fast fashion, greenwashing, and how environmentally conscious consumers build brand confidence. The final cluster, the seventh, focuses on sustainable apparel and clothing design elements. It contains keywords associated with upcycling, principles of sustainable clothing design, and creative methods of producing ecologically friendly apparel.

These clusters facilitate deeper Study and insights into this quickly developing topic by offering an organized framework for comprehending and assessing many facets of sustainability within the fashion business.

Future Research Directions

Future research in sustainable fashion could focus on the intricate interplay among sustainable design and sustainable consumption, purchase intention and luxury brands, fashion and sustainability, supply chain management within the fashion industry, CSR and sustainable development, sustainable fashion and consumer behavior, and sustainable clothing design. Exploring the synergies and tensions between these clusters presents an opportunity to develop a comprehensive understanding of sustainable practices in the fashion sector. Researchers might delve into the nuanced relationships between sustainable design principles and consumer purchasing decisions, particularly in the context of luxury brands. Additionally, investigating the impact of supply chain management on sustainability goals and aligning corporate social responsibility initiatives with broader sustainable development objectives could unveil critical insights. A holistic approach to studying sustainable fashion, encompassing design and consumer behavior, is essential for creating effective strategies that foster environmentally conscious practices throughout the industry.

Implications of the Study

The findings are significant for both academia and industry. Academically, the bibliometric review provides a comprehensive overview of the existing literature on eco-brand expansion strategies and brand engagement within the fashion industry, offering a foundation for future research directions. The identified trends and gaps in literature can guide scholars toward unexplored areas, stimulating further inquiry into the dynamics of sustainable branding in fashion. The Study offers insights into the evolving landscape of eco-branding strategies for industry practitioners, enabling fashion brands to refine their approaches for better consumer engagement. The implications extend to potentially enhancing sustainable practices within the fashion industry to foster sustainable production and consumption.

Findings and Conclusion

The bibliometric review has shown noteworthy patterns and perspectives in the current body of literature, providing insight into how eco-branding tactics are developing and how they affect consumer interaction with brands. A remarkable discovery is the increasing scholarly focus on the relationship between brand engagement and eco-branding strategy, which reflects the fashion industry's expanding concern for sustainability. This ever-increasing interest demonstrates the industry's growing interest in incorporating environmental responsibility into branding initiatives. To be current and competitive, fashion firms must embrace eco-friendly efforts as consumers increasingly prioritize sustainable practices. As a result, a greater focus has been on the relationship between eco-branding tactics and brand engagement and how these factors affect customer behavior and brand perception. Furthermore, the bibliometric analysis clarifies a noticeable upsurge in studies investigating the impact of sustainable design on brand interaction. This upsurge highlights a paradigm shift in the fashion sector wherein eco-friendly product development becomes a brand's key differentiator and a draw for consumers.

The review highlights how important sustainable design is in influencing consumer attitudes and encouraging brand loyalty. Sustainable brands not only meet ethical requirements but also build stronger relationships with their audience as consumers grow increasingly aware of the environmental impact of their purchases. Beyond only creating innovative products, sustainable design becomes a potent force behind advocacy and brand loyalty. Moreover, bibliometric analysis discloses a variety of approaches utilized in researching the connection between sustainable design and brand interaction. Scholars utilize a range of methodologies to decipher the intricacies of this dynamically changing environment, ranging from quantitative evaluations of consumer behavior to qualitative investigations of brand narratives. The bibliometric review highlights the changing conversation around eco-branding tactics and brand interaction in the fashion industry. It highlights a paradigm change in which sustainability is no longer an afterthought but a fundamental component of brand development. As researchers explore this intersection more thoroughly, the results improve scholarly research and provide industry practitioners with crucial information for managing the market's growing sustainability consciousness.

References

Athwal, N., Wells, V. K., Carrigan, M., & Henninger, C. E. (2019). Sustainable luxury marketing: A synthesis and research agenda. International Journal of Management Reviews, 21(4), 405–426.

Bertola, P., & Teunissen, J. (2018). Fashion 4.0. Innovating fashion industry through digital transformation. Research Journal of Textile and Apparel, 22(4), 352–369.

Caniato, F., Caridi, M., Crippa, L., & Moretto, A. (2012). Environmental sustainability in fashion supply chains: An exploratory case based research. International journal of production economics, 135(2), 659–670.

Chadegani, A. A., Salehi, H., Yunus, M. M., Farhadi, H., Fooladi, M., Farhadi, M., et al. (2013). A comparison between two main academic literature collections: web of science and scopus databases. Asian Social Science, 9(5).

Chatterjee, P. (2009). Green brand extension strategy and online communities. Journal of Systems and Information Technology, 11(4), 367–384.

Delafrooz, N., Taleghani, M., and Nouri, B. (2014). Effect of green marketing on consumer purchase behavior. QScience Connect, 2014(1), 5.

Di Leo, A., Sfodera, F., Cucari, N., Mattia, G., and Dezi, L. (2023). Sustainability reporting practices: an explorative analysis of luxury fashion brands. Management Decision, 61(5).

Geegamage, T., Ranaweera, A., and Halwatura, R. (2023). Pre-loved or hatred? consumers' perception of value towards secondhand Fashion consumption in Sri Lanka. Research Journal of Textile and Apparel., 28(2), 91.

Han, J., Seo, Y., & Ko, E. (2017). Staging luxury experiences for understanding sustainable fashion consumption: A balance theory application. Journal of Business Research, 74, 162–167.

Henninger, C. E., Alevizou, P. J., and Oates, C. J. (2016). What is sustainable fashion? Journal of Fashion Marketing and Management: An International Journal, 20(4), 400–416.

Hill, J., and Lee, H. H. (2015). Sustainable brand extensions of fast fashion retailers. Journal of Fashion Marketing and Management, 19(2), 205–222.

Jain, S., and Mishra, S. (2020). Luxury fashion consumption in sharing economy: a study of Indian millennials. Journal of Global Fashion Marketing, 11(2), 171–189.

Kapoor, P. S., Tagore, A., and Dua, S. (2023). Social media influencer promoted sustainable fashion: effects of sponsorship and benefit association. Journal of Promotion Management, 29(4), 461–490.

Ki, C., and Kim, Y. K. (2016). Sustainable luxury fashion consumption and the moderating role of guilt. Fashion, Industry and Education, 14(1), 18–30.

Kim, T., Kim, D., and Park, Y. (2022). Recent progress in regenerated fibers for 'Green' textile products. Journal of Cleaner Production, 376, 134226.

Lee, E. J., Bae, J., and Kim, K. H. (2020). The effect of sustainable certification reputation on consumer behavior in the fashion industry: focusing on the mechanism of congruence. Journal of Global Fashion Marketing, 11(2), 137–153.

Marcella-Hood, M. (2023). Augmenting sustainable fashion on Instagram. Sustainability, 15(4), 3609.

Nagurney, A., & Yu, M. (2012). Sustainable fashion supply chain management under oligopolistic competition and brand differentiation. International Journal of Production Economics, 135(2), 532–540.

Olsen, M. C., Slotegraaf, R. J., and Chandukala, S. R. (2014). Green claims and message frames: how green new products change brand attitude. Journal of Marketing, 78(5), 119–137.

Putri, A. D., and Yeshika, A. (2024). The moderating role of generation differences determinant factors of sustainable apparel behavior intention. Indonesian Journal of Business and Entrepreneurship, 10(1), 154–154.

Retamal, M., Brydges, T., Sharpe, S., Ferrero-Regis, T., Fisher, D., Gwilt, A., et al. (2023). State of play in Australian sustainable fashion research: current and future directions. International Journal of Fashion Studies, 10(1), 19–49.

Sinha, P., Sharma, M., and Agrawal, R. (2023). A systematic review and future research agenda for sustainable fashion in the apparel industry. Benchmarking: An International Journal, 30(9).

Wang, Y., Chowdhury Ahmed, S., Deng, S., and Wang, H. (2019). Success of social media marketing efforts in retaining sustainable online consumers: an empirical analysis on the online fashion retail market. Sustainability, 11(13), 3596.

Yao, Z., Zhang, X., Luo, J., and Huang, H. (2020). Offense is the best defense: the impact of workplace bullying on knowledge hiding. Journal of Knowledge Management, 24(3), 675–695.

4 Gamifying the shopping experience: a systematic literature review on gamification in e-commerce apps

Anand Thakur[a], *Ishita Bansal*[b] *and Mohamed Isham A. R.*[c]
Department of Financial Administration, Central University of Punjab, Bathinda, India

Abstract

Gamification is a relatively new phrase that generally refers to using game features in non-gaming circumstances. However, it is gaining traction in various corporate realms. Firms gradually integrate gamification into their mobile applications (apps) and online platforms to optimize their consumers' digital interactions. While numerous systematic studies have evaluated the influence of gamification on customer involvement in other disciplines, they have yet to examine the impact of its utilization in e-commerce apps. This research seeks to explore empirical studies analyzing the function of gamification in e-commerce mobile apps using a systematic literature review. A final pool of 32 empirical research articles published in Scopus-indexed journals was identified. Findings revealed five notable customer behavior outcomes: customer loyalty, consumer engagement, immersive experience, purchase intention, and customer retention. Finally, significant research directions for future relevant investigations have been suggested. As businesses embrace digital innovation, more research in this domain will be required to understand the changing dynamics of consumer behavior in the digital marketplace and to develop effective strategies to capitalize on gamification's benefits for long-term business growth and competitive advantage. This study reveals the revolutionary potential of gamification in reshaping consumer behavior paradigms and also leads the way in exploring gamification in e-commerce mobile apps.

Keywords: Consumer behavior, E-commerce, gamification, mobile applications, systematic literature review

Introduction

Embedding game mechanics into non-gaming environments is a recent trend aimed at motivating people to alter their behavior (Xu et al., 2023). Marketing involves manipulation, persuasion, and motivation, and gamification has much potential to offer (Sigala, 2015. 189). Till 2020, the total amount spent worldwide on gamification solutions was $11.1 billion (Eppmann et al., 2018). Nonetheless, recent projections indicate that the gamification industry will expand at a 'compound annual growth rate' (CAGR) of 0.25%, from USD 57.99 billion in 2023 to USD 58.71 billion by 2028 (Mordor Intelligence, 2024). Gamification emerges as a significant marketing strategy, transforming how companies interact with customers (Xu et al., 2023). Gamification, in its broadest definition, refers to technical, economic, cultural, and sociological advances that fill reality with game-like features, intentionally or as a result of emergent change. (Koivisto and Hamari 2019). Gamification incorporates game design aspects into non-gaming environments to make mundane work pleasurable and exciting (Raman, 2021, 388). The work processes are redesigned with game mechanics to provide a pleasant and entertaining

[a]anand.thakur@cup.edu.in, [b]bansalishita794@gmail.com, [c]ishamohamed333@gmail.com

DOI: 10.1201/9781003564447-4

experience. Also, it is a combination of different elements that create a sequence of desired mechanics that produce interaction behavior and dynamics and assist important business activities (Wood and Reiners, 2015).

Many prosperous firms are concentrating their whole offering on gamifying a fundamental activity (Hamari et al., 2014). In retail, high-end labels such as Nike collaborate with NFT manufacturer 'RTFKT' to create digital trainers for AVATAR. At the same time, Adidas has partnered with 'Bored Ape Yacht Club' to create personalized digitally wearable apparel AVATARS that they sell to customers (Arya et al., 2024). Amazon India and Myntra used spinning wheels, surprise coupons, sign-up incentives, discount vouchers via scratch cards, and loyalty programs (Spais et al., 2022). Likewise, there has been a growing application of game-based therapies in the medical field to encourage motivation, engagement, and long-term sustainability of healthy behaviors (Damaševičius et al., 2023). In the banking and finance industry, the opportunity for gamification features will make a massive difference in the digital banking scenario (Çera et al., 2020). Some banks encourage completing financial chores, such as moving funds to a savings account or investment portfolio, by creating game-like settings in their web and mobile apps. The user advances to a new level or receives points for each activity completed (McNamee, 2023). These instances showcase how the gamified design in branded applications (apps) helps increase consumer engagement with the company (Lin et al., 2022).

Gamification in marketing is a successful approach to increasing sales and meeting corporate objectives across devices (McDonald, 2023). Since customers have more options in the virtual world, businesses must focus on holding their attention. "Gamification," a term that has recently gained popularity, is one way to keep users interested and encourage them to spend more time on an app. This becomes especially important in e-commerce, where the primary shopping experience has historically been transactional. By incorporating game mechanics into the purchasing process, e-commerce players can change this perception and create a more enjoyable and engaging experience (Raman, 2021, 388). Gamification is becoming a potent technique for conversion as the number of mobile customers keeps growing and e-commerce takes a more significant share of the over $22 trillion global retail sector. Gamification leverages engaging game design components to turn users of online apps into devoted patrons. Hence, the current research seeks to review the studies conducted in the context of gamified e-commerce apps and answer the research question: How do gamified elements in e-commerce mobile apps influence customer behavior towards these apps? Though there have been systematic reviews on gamification (Silva et al., 2019), the field offers a further scope of enquiry.

Methodology

This study uses a systematic review process following the 'Scientific Procedures and Rationales for Systematic Literature Reviews (SPAR-4-SLR)' framework (Paul et al., 2021), which includes developing review questions such as, "How do gamified elements in e-commerce mobile apps influence customer behavior towards these apps?". This is a comprehensive framework that researchers may depend on to direct and support their conclusions (Paul et al., 2021). Figure 4.1 depicts the steps that comprise the establishment of the parameters for inclusion and exclusion, the formulation of a search plan, the identification and choosing of studies, and the evaluation and interpretation of data. Scopus electronic databases were utilized to research the studies. This study reviews empirical

Figure 4.1 Flow chart for selection of relevant studies
Source: Authors Compailation

articles on the relationship between gamification elements and consumer behavior, published in reputable scholarly publications from 2011 to 2023. The duration of ten years was selected as the trend in gamification studies about consumer behavior increased after 2011. This year, a gamification summit in San Fransisco was held, which brought together researchers and students worldwide (Cloke, 2023). The initial search with keywords like gamification, consumer behavior, E-commerce, gamification design, gamification elements, and game mechanics yielded 89 studies. Then, after narrowing the subject area to finance, accounting, and business studies, document type to articles, and language to English, 70 studies were chosen for further review. The articles' content was then analyzed by cross-checking their abstracts to verify that the database only returned relevant articles for examination. This approach resulted in the omission of 38 articles. After the elimination procedure and full-text evaluation, 32 publications were chosen for this investigation. Using Microsoft Excel, each article was classified according to the following criteria: the name of the source, the author(s), the document title, the publication year, the theory applied, the technique utilized, the context, and the dependent and independent variables.

Review of Literature

Hsu and Chen (2018a) examined the causes and effects of user experience within online gamification. Researchers found that both utilitarian and hedonistic characteristics are influenced by perceived mobility and that perceived mobility and these features affect user experience. User experience affects brand equity, various types of perceived value, and perceived benefits. The research highlights the mediator role of user experience, providing significant perspectives for website administrators.

Hsu (2023) assessed the relationship among 'brand identity', 'self-expression', 'brand love', 'gamification', 'consumer engagement', 'brand experience', and 'repurchase intention' toward gamified running software Nike Run Club. The study proved that gamification significantly affects brand experience, repurchase intention, and brand love. Also, gamification significantly improves these crucial consumer-brand interactions by moderating the relationship between brand love and customer engagement.

Lin et al. (2022) examined the effect of gamification components (success, challenge, and rewards) on customer engagement, perceived playfulness, attitude, and behavioral intention towards the Starbucks app using the Mechanics Dynamics Emotions and the Technology Acceptance Model. Findings showed that gamification increases user engagement within the app and boosts user satisfaction and loyalty.

Xi and Hamari (2019, 216) examined the impact of gamification features ('immersion', 'achievement', and 'social-related features') on the satisfaction of intrinsic needs ('autonomy', 'competence', and 'relatedness needs'). The findings revealed that while achievement-related characteristics are the best predictor of autonomy and competence needs fulfilment, immersion-related variables favor autonomy needs satisfaction. Socially related characteristics positively impact autonomy, competence, and relatedness need satisfaction.

Hsu and Chen (2018b, 124) assessed the influence of gamification marketing activities (GMAs) on consumer behavior in internet retailing platforms. The research model investigated the links between GMAs' experience, value (hedonic and utilitarian), satisfaction, brand love, and desirable consumer behaviors. The findings highlight that experiences of GMA impact hedonic and utilitarian values intensely and favorably. Furthermore, hedonic and utilitarian factors highly influence satisfaction and brand love.

Bitrián et al. (2020) determined how game components in mobile sports apps, like Nike Plus or Strava, affect users' autonomous motivation and satisfaction with basic psychological requirements, such as competence, autonomy, and relatedness. Using self-determination theory (SDT), the study discovered that immersion-related game features satiate demands for competence and autonomy, achievement-related game elements improve all three psychological needs, and social game components accomplish the requirement for connection. Moreover, the study underscores the importance of fulfilling relatedness and autonomy needs during app interactions to foster autonomous motivation.

Al-Zyoud (2021) supported the idea that gamification increases customer loyalty, encourages purchase intention, and promotes positive word-of-mouth (WOM). Consumer loyalty influenced both purchase intention and e-WOM sharing behavior. Additionally, it acted as a mediator between gamification and these behaviors.

Torres et al. (2022) explored how gamification affects brand loyalty and WOM. Authors found that hedonic and social values strongly impact consumer behavior. However, they also identified utilitarian value contributing to positive WOM and brand loyalty. Further, brand love mediates brand loyalty more so than WOM. Different approaches may be needed to achieve desired results, considering varying circumstances. Integrating utilitarian and social values with satisfaction and brand love drives success in both outcomes.

Jang and Hsieh (2021) discovered that gamification and virtual reality (VR) optimized online services can impact customer preferences in tourist e-commerce. The authors demonstrated that gamification enjoyment and activation positively affected media richness, substantially impacting the utility and convenience of utilizing VR technology with gamification. Additionally, utility, usability, engagement, and immersion positively impact users' perception of a gamified and VR-enhanced travel system. Furthermore, presence was shown to have a favorable effect on immersion. Finally, the user's perceived value influenced the intention to adopt via a positive effect on satisfaction.

Kim et al. (2020) discovered that hedonic value, preceded by gamification, positively influences repurchase intention. Furthermore, the study model suggests that novelty-seeking has a moderating influence on repurchase intention.

Kaur et al. (2023) assessed how game apps influence a consumer's purchase intention when perceived satisfaction, hedonic value, and social connection are mediators. The study uses the mechanics-dynamics-aesthetics (MDA) framework. The findings suggested that fun, as an element of mobile gamification, might significantly impact social interaction, and storytelling significantly impacts perceived enjoyment. Lastly, perceived enjoyment significantly predicts customer intent to use gamification.

Using goal-setting theory, Cao et al. (2022) explored the effect of gamified encounters, particularly competitiveness and collaboration, on customers' carbon-free behaviors. Both impact the perceived environmental responsibility, perceived enjoyment, and social gain, which will impact low-carbon actions. Drawing from survey data from 'Ant Forest' users on 'Alipay', both interactions promote low-carbon actions, with cooperation having a more significant effect. Cooperative interactions stimulate normative, hedonic, and gain motivations, while competitive interactions only motivate hedonic and gain goals.

Luna-Perejon et al. (2019) evaluated the performance of a smartphone app designed to help people quit smoking in terms of usability, satisfaction, and user experience. There were two user groups involved: experts and smokers. The app features included benefits, diversionary minigames, and inspirational quotes. The results of expert assessments showed an 80% satisfaction score for motivational messages and over 69% agreement on game playability.

Braga et al. (2023) explored the influence of game components and varying budgets on participants' dietary choices. The findings demonstrated that those exposed to game components selected food bundles with higher nutritional value, underscoring the potential of gamification to encourage healthy online grocery shopping practices.

Pamuru et al. (2021) explored how local restaurants are affected financially by the augmented reality (AR) mobile game Pokémon Go. The research approaches the game's release as a natural experiment, using online reviews to gauge customer interest. The results show that eateries connected to Pokémon have more customer satisfaction and favorable perceptions.

Li (2018) examined consumers' switching between branded apps and regular membership cards. Push effects like 'poor aesthetics, mooring effects like perceived substitutability and inertia, and pull effects like capability, transaction convenience, economic benefit, and gamification' impact the consumer's switching intention. The findings imply that switching intention is positively influenced by push factors like 'aesthetics and pull factors like sociability, transaction convenience, economic rewards, and gamification'. Furthermore, perceived substitutability and inertia moderate the link between pull, push, and switching intention.

Kunkel et al. (2023) investigated how varying motivating affordances included in alerts influence the use of the app afterwards. The results showed that notifications with progression incentives caused the most significant increases in user behavior. Progression alone was less effective than social incentives or a mix of social and progression affordances. The degree of consumer interaction with the brand increased its influence.

Tan et al. (2017) examined how technological affordances affect the disruptive characteristics of multi-sided digital platforms, with a particular emphasis on the Australian e-hailing platform Go-Catch. Findings reveal that situational affordance, artefactual affordance and actualization impact economic access. The study proposes a model that shows how technology affects consumer recognition, engagement, and transposition.

Sigala (2015) examined the application of gamification and its effects on TripAdvisor's fun ware in a particular tourism scenario. Findings indicated three fundamental factors of apps' process, outcome, and social interaction factors' impact—customers' perception of the gamified platforms and their behavioral intention.

Parapanos and Michopoulou (2023) explored the evolving hospitality business due to increased mobile technology use, and the research explores what 'fun' means to guests using mobile hotel gamified apps. Fun typology, socializing, exploring, achieving, challenge, competitiveness, interactivity and personalization are the variables that accelerate the use of technology among hotel app users.

Wang et al. (2018) analyzed the effects of gamification on user-generated reviews, particularly WOM contributions. Gamification components like badges impact perceived competence and review usefulness, which drives the consumers' behavioral outcomes. Drawing on research in social psychology, it proposes that positive perceptions of WOM contributors' competence are enhanced by gamification symbols signaling experiences, which alters behavioral outcomes like propensity to visit/buy and recommend reviewed outlets.

Bauer et al. (2020) investigated the impact of combining games and monetary prizes on consumer happiness, loyalty, and WOM intentions in online retail. Findings suggest that Shopping-related games lacking financial benefits improve all three relational outcomes due to increased customer satisfaction with the overall shopping experience. However, using monetary incentives to promote game participation diminishes these effects.

Zhang et al. (2021a) investigated how gamification influences customers' impulsive purchase decisions. The study develops a theoretical framework to analyze how two gamification processes, badge upgrading and reward giving, influence social interaction, perceived enjoyment, and impulsive purchases. The findings suggest that both gamification tactics impact customers' impulsive purchases through social interaction and perceived enjoyment.

Zhang et al. (2021b) explored how well gamification techniques like competitive leaderboards might increase people's intentions to save. The outcomes exhibit that while social comparison orientation has a discernible impact on the effect, leaderboards significantly impact saving intentions when people compare themselves to higher standards.

Wut et al. (2021) advocated adopting the extended 'Unified Theory of Acceptance and Use of Technology' (UTAUT) model within gamification and the design and attributes of consumption-related mobile apps. The study found that gamification characteristics positively affect user behavior. Extended UTAUT model components (performance expectancy, effort expectancy, facilitating conditions) and app design determine behavioral intentions to use gamified mobile apps among young consumers.

Xu et al. (2023) indicated that gamification affordances, such as fun, monetary rewardability, and novelty, had a favorable impact on user immersion. Further analysis has demonstrated a significant connection between the level of immersion experienced by users and their intent to make purchases within a gamified environment.

Xu et al. (2022) used the 'Affordances-Psychological Outcomes-Behavioral Outcomes' paradigm to explore how perceived gamification features influence psychological ownership and consumer citizenship behaviors (CCB). Authors found that autonomy, self-expression, interactivity, and competition impact psychological ownership, which in turn supports CCB. Autonomy and interactivity indirectly influence CCB through psychological ownership. Control variables like usage frequency, age, gender, and social influence also affect CCB.

Zhang et al. (2023) showed that user engagement is favorably associated with four gamified artefactual affordances ('rewards, competition, feedback, and cooperation'), improving the retention of users on platforms. User engagement is associated with retention on mobile payment platforms. In addition, the effect of rewards on user retention is mediated by user engagement, whereas competition, feedback, and cooperative affordances are partially mediated. The effect of four gamified artefactual affordances on user engagement is moderated by narrative affordance.

Shi et al. (2022) used affordance theory and consumption values to discover the OTA platform's gamification affordances throughout online purchasing events. Mixed method studies were conducted to determine how these affordances impact customers' value perceptions and purchase intentions. Findings revealed that gamification affordances ('achievement, identity, self-expression, competition') improve consumer value. Also, utilitarian and emotional values positively impact consumers' purchase intentions.

Cicchirillo (2024) studied consumers' gratifications with gamified e-services, which drives their continued usage. The author found that entertainment, convenience, and economic gratifications were favorably related to attitudes towards a gamified service and, indirectly, with continuing usage via attitudes. Gamified services that are challenging generate pleasant emotions and encourage users to keep using them, while social and informational rewards do not predict users' attitudes or future usage intention.

Alvi (2022, 217) used the SDT to examine how gamification in language learning apps affects users' perceptions of brand engagement and consumer behavior. Absorption and social elements-related interactions do not affect consumer engagement. Further, brand engagement is associated with relationship quality factors like satisfaction, trust and commitment. Lastly, trust increased commitment but not satisfaction among app users.

De Canio et al. (2021) indicated that shopping gamification is crucial for customers to engage with mobile apps. Further, the authors examine the impact of additional intrinsic incentives on the engagement-intention-to-buy relationship. Findings reveal a favorable association between socialness, perceived enjoyment, focused attention and consumer engagement. Results also confirm the moderating role of previous shopping experiences between engagement and buying intention.

Analysis and Results

Source: The Table 4.1 illustrates the diverse sources of empirical articles selected for the review, showcasing a broad range of journals contributing to the research in gamification.

Country: Table 4.2 depicts the geographical emphasis of gamification studies. China, Taiwan, and the United States were the primary focus of gamification studies.

Context: Table 4.3 presents the platforms and markets investigated. Most studies were conducted in the e-commerce, hospitality, and tourism sectors.

Theory: Table 4.4 presents the theories that form the basis for model building and conceptualization in gamification research. The diversity of theories reflects the multidisciplinary nature of gamification research, drawing from the disciplines of psychology and information systems to

Table 4.1 Sources of empirical articles selected for SLR.

Name of journal	TP	Name of journal	TP
International Journal of Information Management	3	Information Systems Research,	1
Decision Support Systems	2	Electronic Markets	1
Computers in Human Behavior	2	Journal of Content, Community and Communication	1
Journal of Environmental Management	1	Tourism Planning & Development	1
PloS one	1	Frontiers in Psychology	1
Computer Methods and Programs in Biomedicine	1	Journal of Global Information Management	1
Sustainability	1	Journal of Theoretical and Applied Electronic Commerce Research	1
Food Quality and Preference	1	European Journal of Marketing	1
Technological Forecasting and Social Change	1	Multimedia Tools and Applications	1
Management Information Systems Quarterly	1	Psychology & Marketing	1
Journal of Service Management	1	Young Consumers	1
European Journal of Management and Business Economics	1	Tourism Management	1
Journal of Public Affairs	1	Services Marketing Quarterly	1
Communications of the Association for Information Systems	1	International Journal of Retail & Distribution Management	1

Note: TP-Total publications

Source: Authors' compilation

Table 4.2 Countries.

Country	Authors
China	Xu et al. (2023); Zhang et al. (2023); Cao et al. (2022); Xu et al. (2022); Zhang et al. (2021a); Shi et al. (2022); De Canio et al. (2021); Xi and Hamari (2019, 211).
Taiwan	Hsu (2023, 4); Luna-Perejon et al. (2019); Lin et al. (2022); Hsu and Chen (2018b, 127); Li (2018, 177); Jang and Hsieh (2021, 29350)
United States	Braga et al. (2023); Zhang et al. (2021b); Kunkel et al. (2023); Pamuru et al. (2021); Wang et al. (2018); Bauer et al. (2020)

Source: Authors' compilation

Table 4.3 Context of gamification studies.

Context (Platforms)	Authors
E-retail	Zhang et al. (2021a); Xu et al. (2023); Xu et al. (2022); De Canio et al. (2021); Kim et al. (2020); Braga et al. (2023); Hsu and Chen (2018b, 127); Bauer et al. (2020); Al-Zyoud (2021, 3)
Tourism and hospitality	Lin et al. (2022). Li (2018, 177); Jang and Hsieh (2021, 29350); Wang et al. (2018); Li (2018); Parapanos and Michopoulo (2023, 169); Shi et al. (2022); Cicchirillo (2024, 11); Sigala (2015, 189); Pamuru et al. (2021)

Source: Authors' compilation

Table 4.4 Theories used in gamification studies.

Theory	Authors
Self-determination theory	Xi and Hamari (2019, 211); Bitrián et al. (2020); Sigala, (2015, 191); Alvi (2022, 211); Kunkel et al. (2023)
Goal-setting theory	Cao et al. (2022)
MDA framework	Lin et al. (2022); Kaur et al. (2023)
Means-end chain theory	Kim et al. (2020)
Technology adoption model	De Canio et al. (2021)
Uses and gratification theory	Wut et al. (2021); Cicchirillo (2024, 3)
UTAUT Model	Wut et al., 2021
affordance theory	Xu et al. (2023); Kunkel et al. (2023); Zhang et al. (2023); Xu et al. (2022); Shi et al. (2022); Tan et al. (2017).
Social cognition theory	Wang et al. (2018)
Social comparison theory	Zhang et al. (2021b); Kunkel et al. (2023)
Situation relevance theory	Sigala (2015, 191)
Push-pull mooring framework	Li (2018, 172)
Media richness theory	Jang and Hsieh (2021, 29342)
Theory of consumption values	Shi et al. (2022)

Source: Authors' compilation

Table 4.5 Widely used methods in gamification studies.

Surveys	Xu et al. (2023); Cicchirillo (2024, 11); Kaur et al. (2023); Hsu (2023, 4); Xu et al. (2022); Lin et al. (2022); Torres et al. (2022); Cao et al. (2022); Alvi (2022, 215); Zhang et al. (2021a); Wut et al. (2021); De Canio et al. (2021); Jang and Hsieh (2021, 29350); Al-Zyoud (2021, 3); Bitrián et al. (2020); Xi and Hamari (2019, 213); Li (2018, 177); Hsu and Chen (2018b, 126); Hsu and Chen (2018a, 6); Sigala (2015, 200).
Experiments	Braga et al. (2023); Wang et al. (2018); Bauer et al. (2020); Pamuru et al. (2021)
Qualitative studies	Luna-Perejon et al. (2019); Tan et al. (2017); Parapanos and Michopoulo (2023, 168)
Mixed method research	Zhang et al. (2021b); Kunkel et al. (2023); Zhang et al. (2023); Shi et al. (2022); Kim et al. (2020)

Source: Authors' Compilation

understand behavioral dynamics associated with gamification.

Methodology: Table 4.5 provides an overview of typical research techniques used in gamification studies, such as experiments, surveys, mixed-method approaches, and qualitative investigations. Surveys are the most common tools, followed by experiments and mixed methodologies, with few using qualitative research.

Constructs

Table 4.6 lists common independent variables in gamification studies: cooperation,

Table 4.6 Independent variables.

Cooperation	Cao et al. (2022); Zhang et al. (2023)
Competition	Cao et al. (2022); Zhang et al. (2023); Shi et al. (2022); Zhang et al. (2021b); Xu et al. (2022)
Rewards	Xu et al. (2023); Zhang et al. (2023); Kaur et al. (2023); Kunkel et al. (2023); Hsu (2023, 6); Cicchirillo (2024, 15); Lin et al. (2022); Zhang et al. (2021a); Bauer et al. (2020).
Achievement	Lin et al. (2022); Xi and Hamari (2019, 217); Bitrián et al. (2020); Shi et al. (2022); Alvi (2022, 217)
Challenge	Lin et al. (2022)
Goals	Kim et al. (2020); Hsu (2023, 6)
Control	Hsu (2023, 6); Xu et al. (2022)
Immersion	Cicchirillo (2024, 15); Alvi (2022, 217); Bitrián et al. (2020); Xi and Hamari (2019, 217).
Playfulness	Xu, et al., (2023)
Novelty	Xu et al. (2023); Kim et al. (2020); Hsu and Chen (2018b, 129).
Storytelling	Zhang et al. (2023); Xu et al. (2022)
Identity	Shi et al. (2022)
Self-expression	Xu et al. (2022); Shi et al. (2022)
Entertainment	Cicchirillo (2024, 15); Hsu and Chen (2018b, 129)
Prompts, progress tracking, points, badges, leaderboards	Braga et al. (2023); Kaur et al. (2023); Hsu (2023, 6); Zhang et al. (2021b); Zhang et al. (2021a); Wang et al. (2018); Sigala (2015, 201).
Social features	Xi and Hamari (2019, 217); Zhang et al. (2021b); Bitrián et al. (2020); Kunkel et al. (2023); Torres et al. (2022); Xu et al. (2022); Cicchirillo (2024, 15); Alvi (2022, 217); Kim et al. (2020); Hsu (2023, 6); Hsu and Chen, (2018b, 129); Xu et al. (2023); Zhang et al. (2023)

Source: Authors' compilation

competition, rewards, achievement, and social features. Table 4.7 details frequently studied dependent variables such as immersive experience, customer engagement, and intention to purchase. The studies shed light on the relationships between gamification components and psychological needs satisfaction, emphasizing the importance of tailored gamification strategies. Game features, encompassing success, challenge, and rewards, emerge as a catalyst for heightened customer engagement, perceived playfulness, and positive attitudes towards apps, ultimately influencing behavioral intentions. Creating a game-like experience may enhance brand engagement, loyalty, and purchase intention.

Discussion and Implications

The main aim of the reviewed studies was to examine how gamification influences different facets of consumer behavior toward mobile apps. This section discusses the varied effects of gamification components on customer behavior.

a) Immersive Experience

By adding components like challenges, badges, and prizes, app gamification improves user experience and encourages motivation. Achievement, rivalry, and advancement provide positive emotional experiences for a more engaging and pleasurable connection. By appealing to users' innate motives, these game-like elements

Table 4.7 Dependent variables.

Constructs	Authors
Immersive experience	Hsu and Chen (2018a, 9); Hsu (2023, 6); Braga et al. (2023); Sigala (2015, 204); Xu et al. (2023); Jang and Hsieh (2021, 29355); Bauer et al. (2020)
Customer engagement	Zhang et al. (2023); Hsu (2023, 6); Lin et al. (2022); Alvi (2022, 217); Tan et al. (2017).
Customer loyalty	Hsu and Chen (2018a, 9); Hsu and Chen (2018b, 129); Al-Zyoud (2021, 5); Torres et al. (2022); Bauer et al. (2020)
Intention to purchase	Lin et al. (2022); Zhang et al. (2021b); Sigala (2015, 190); Kaur et al. (2023); Xu et al. (2023); Jang and Hsieh, (2021, 29355); Wut et al. (2021); Shi et al. (2022); De Canio et al. (2021)
Customer retention	Hsu (2023, 6); Zhang et al. (2023); Cicchirillo (2024, 15); Kim et al. (2020)

Source: Authors' Compilation

enhance the app's user experience and provide them with an emotionally fulfilling encounter. Additionally, well-designed incentive systems in retail games provide players with various chances to engage with the activities fully, leading to an immersive experience. Gamification elements impact intrinsic need satisfaction, leading to a feeling of absorption and immersion in the purchase environment (Xu et al., 2023).

b) Customer Engagement

App gamification uses game-like components to impact user behavior positively. Integrating interactive challenges, badges, and rewards improves user pleasure, motivation, and engagement. This strategy creates a sense of accomplishment by appealing to our innate need for success and acknowledgement. As a result, users are more likely to remain loyal, take up desired habits, and use the app more regularly, all contributing to a vibrant and entertaining user experience. Gamification impacts consumers' cognitive and emotional engagement with platforms by stimulating curiosity (Zhang et al., 2023). Moreover, gamification solutions address users' desires for enhanced immersion, including entertainment, social interaction, and monetary rewards, fostering sustained engagement (Zhang et al., 2023).

c) Customer Loyalty

App gamification increases loyalty by encouraging engagement through rewards, social interaction, and personalized experiences. Rewards, badges, and scores provide a sense of achievement, while social components and competition develop a sense of cohesiveness and improve customer loyalty. Personalized content and incentive-based loyalty programs support continued user engagement. GMA experience benefits the emotional and functional benefits derived, influencing consumer satisfaction and brand love. Brand love enhances desirable consumer behavior through brand loyalty and positive word of mouth (Hsu and Chen, 2018a, 8; Torres et al., 2022). Gamification enhances users' emotional, cognitive and social engagement, resulting in customer loyalty (Xi and Hamari, 2019, 216). Games elicit strong emotional and motivational responses in players, which can lead to a strong connection to the surroundings and the games themselves.

d) Intention to Purchase

App gamification enhances customer experience through fun, prizes, storytelling, and other mechanics that ultimately promote social interaction, perceived enjoyment and hedonic value, all of which have a favorable impact on users' intention to buy and use the app (Kaur et al., 2023). Gamification improves the user experience by adding competitive features and personalized experiences, encouraging users to spend time and money on the app. In-app purchases, learning opportunities, and a sense of accomplishment result in an engaging and pleasurable experience for users, raising the possibility of consistent usage and conversions. Gamification fulfils the psychological requirements of consumers and promotes online purchase intention (Xu et al., 2022).

e) Customer Retention

By incorporating game-like elements like leaderboards and awards, app gamification increases user retention and discourages switching. Apps provide a feeling of community and brand loyalty by encouraging social interactions and offering incentives for behaviors. While gamification improves user experience and helps apps stand out in crowded markets, its effectiveness depends on careful planning and continuous innovation to keep users engaged and avoid getting tired of using the app. Gamification creates emotional connections by leveraging intrinsic motivators like competition and success. This increases the likelihood that users will return to the app because of its delight and satisfaction. Gamified app designers may target value-in-behavior to develop experiences that motivate users to repeat desirable behaviors and use them. (Torres et al., 2022). When customers are drawn in by the prospect of further benefits like financial savings, it enhances customer satisfaction and retention.

Theoretical and Managerial Implications

The research has theoretical significance since it carefully reviews the gamification tactics used in e-commerce apps, enhancing the knowledge of the relationship between gamification and online buying. The investigation contributes to consumer behavior theories by investigating how gamification features impact decision-making in e-commerce. This might result in modifying or expanding current models to account for the distinct dynamics of gamification in digital buying environments. Furthermore, the review may reveal the psychological factors that underpin customer engagement and pleasure, improving the understanding of motivation, reward systems, and emotional connections in the e-commerce area.

The article has significant managerial ramifications for e-commerce enterprises. It gives valuable information for the strategic design of e-commerce apps. It encourages companies to use successful gamification to improve the buying experience, potentially affecting consumer opinions and driving repeat app usage. Furthermore, the study may help organizations design customized marketing strategies for e-commerce apps, employing gamification concepts to boost user engagement and sales. Insights on user retention and loyalty programs may help firms enhance their tactics, cultivating client loyalty through engaging and rewarding gamified experiences. Furthermore, the study might help organizations improve product suggestions and upselling techniques by including gamified components, making them more engaging and productive.

Conclusion

In conclusion, this review examines how gamification affects consumer behavior to comprehend its dynamic impact on user engagement, experience, and loyalty suggests future research avenues. The results show how powerful gamification can be, with components like achievement, challenge, and incentives impacting customer engagement, loyalty, and satisfying intrinsic needs across various industries. The research highlights the adaptability and potential of gamification, adding to a growing understanding of this interactive marketing tool.

Further research in this ever-evolving sector will guide future marketing strategies.

Acknowledgement

The authors gratefully acknowledge the Central University of Punjab for its support in the research.

References

Alvi, I. (2022). Gamification interaction features in language learning apps, brand engagement, and brand-consumer relationship quality. *Journal of Content, Community and Communication*, 15(8), 210–226.

Al-Zyoud, M. F. (2021). The impact of gamification on consumer loyalty, electronic word-of-mouth sharing and purchase behaviour. *Journal of Public Affairs*, 21(3), e2263.

Arya, V., Sambyal, R., Sharma, A., and Dwivedi, Y. K. (2024). Brands are calling your AVATAR in metaverse–a study to explore XR-based gamification marketing activities & consumer-based brand equity in the virtual world. *Journal of Consumer Behaviour*, 23(2), 556–585.

Bauer, J. C., Linzmajer, M., Nagengast, L., Rudolph, T., and D'Cruz, E. (2020). Gamifying the digital shopping experience: games without monetary participation incentives increase customer satisfaction and loyalty. *Journal of Service Management*, 31(3), 563–595.

Bitrián, P., Buil, I., and Catalán, S. (2020). Gamification in sports apps: the determinants of users' motivation. *European Journal of Management and Business Economics*, 29(3), 365–381.

Braga, B. C., Cash, S. B., Sarson, K., Chang, R., Mosca, A., and Wilson, N. L. W. (2023). The creation of an online grocery store for experimental purposes: a pilot study. *Food Quality and Preference,* 109, 104888.

Cao, Y., Kou, F., Hu, H., and Wan, G. (2022). How gamified cooperation and competition motivate low-carbon actions: an investigation of gamification in a popular online payment platform in China. *Journal of Environmental Management*, 324, 116259.

Çera, G., Pagria, I., Khan, K. A., and Muaremi, L. (2020). Mobile banking usage and gamification: the moderating effect of generational cohorts. *Journal of Systems and Information Technology*, 22(3), 243–263.

Cicchirillo, V. (2024). Consumer gratifications for gamified e-services. *Services Marketing Quarterly*, 45(1), 105–127.

Cloke, H. (2023). The history of gamification (From the very beginning to now). Retrieved December 2, 2023, from Growth Engineering website: https://www.growthengineering.co.uk/history-of-gamification/

Damaševičius, R., Maskeliūnas, R., and Blažauskas, T. (2023). Serious games and gamification in healthcare: a meta-review. *Information*, 14(2), 105.

De Canio, F., Fuentes-Blasco, M., and Martinelli, E. (2021). Engaging shoppers through mobile apps: the role of gamification. *International Journal of Retail and Distribution Management*, 49(7), 919–940.

Eppmann, R., Bekk, M., and Klein, K. (2018). Gameful experience in gamification: construction and validation of a gameful experience scale [GAMEX]. *Journal of Interactive Marketing*, 43(1), 98–115.

Hamari, J., Koivisto, J., and Sarsa, H. (2014). Does gamification work? --a literature review of empirical studies on gamification. In 2014, 47th Hawaii International Conference on System Sciences, (pp. 3025–3034). IEEE.

Hsu, C.-L. (2023). Enhancing brand love, customer engagement, brand experience, and repurchase intention: focusing on the role of gamification in mobile apps. *Decision Support Systems*, 174, 114020.

Hsu, C.-L., and Chen, M. C. (2018a). How does gamification improve user experience? an empirical investigation on the antecedences and consequences of user experience and its mediating role. *Technological Forecasting and Social Change,* 132, 118–129.

Hsu, C.-L., and Chen, M. C. (2018b). How gamification marketing activities motivate desirable consumer behaviours: focusing on the role of brand love. *Computers in Human Behavior,* 88, 121–133.

Jang, Y.-T., and Hsieh, P. S. (2021). Understanding consumer behaviour in the multimedia context: incorporating gamification in VR-enhanced web system for tourism e-commerce. *Multimedia Tools and Applications,* 80(19), 29339–29365.

Kaur, J., Lavuri, R., Parida, R., and Singh, S. V. (2023). Exploring the impact of gamification elements in brand apps on the purchase intention of consumers. *Journal of Global Information Management*, 31(1), 1–30.

Kim, C., Costello, F. J., and Lee, K. C. (2020). The unobserved heterogeneous influence of gamification and novelty-seeking traits on consumers' repurchase intention in the omnichannel retailing. *Frontiers in Psychology*, 11, 557144.

Koivisto, J., and Hamari, J. (2019). The rise of motivational information systems: a review of gamification research. *International Journal of Information Management*, 45, 191–210.

Kunkel, T., Hayduk, T., and Lock, D. (2023). Push it real good: the effects of push notifications promoting motivational affordances on consumer behaviour in a gamified mobile app. *European Journal of Marketing*, 57(9), 2592–2618.

Li, C.-Y. (2018). Consumer behaviour in switching between membership cards and mobile applications: The case of Starbucks. *Computers in Human Behavior*, 84, 171–184.

Lin, C.-W., Chien, C. Y., Ou Yang, C. P., and Mao, T. Y. (2022). Encouraging sustainable consumption through gamification in a branded app: a study on consumers' behavioral perspective. *Sustainability*, 15(1), 589.

Luna-Perejon, F., Malwade, S., Styliadis, C., Civit, J., Cascado-Caballero, D., Konstantinidis, E., et al. (2019). Evaluation of user satisfaction and usability of a mobile app for smoking cessation. *Computer Methods and Programs in Biomedicine,* 182, 105042.

McDonald, S. (2023). What is ecommerce gamification? a complete guide. Retrieved December 3, 2023, from Brame website: https://brame.io/blog/ecommerce-gamification/

McNamee, J. (2023). Generative AI has the power to transform the banking industry over the next three years. EMarketer.com. Retrieved April, 27, 2024. https://www.emarketer.com/content/generative-ai-transform-banking

Mordorintelligence.com (2024). Gamification Market Size & Share Analysis - Growth Trends & Forecasts (2024 - 2029) Retrieved from. https://www.mordorintelligence.com/industry-reports/gamification-market

Pamuru, V., Khern-am-nuai, W., and Kannan, K. (2021). The impact of an augmented-reality game on local businesses: a study of Pokémon Go on restaurants. *Information Systems Research*, 32(3), 950–966.

Parapanos, D., and Michopoulou, E. (2023). Innovative mobile technology in hotels and the use of gamification. *Tourism Planning and Development*, 20(2), 162–187.

Paul, J., Lim, W. M., O'Cass, A., Wei Hao, A., and Bresciani, S. (2021). Scientific procedures and rationales for systematic literature reviews (SPAR-4-SLR). *International Journal of Consumer Studies*, 45(4), O1–O16.

Raman, P. (2021). Examining the importance of gamification, social interaction and perceived enjoyment among young female online buyers in India. *Young Consumers*, 22(3), 387–412.

Shi, S., Leung, W. K. S., and Munelli, F. (2022). Gamification in OTA platforms: a mixed-methods research involving online shopping carnival. *Tourism Management*, 88, 104426.

Sigala, M. (2015). The application and impact of gamification funware on trip planning and experiences: the case of TripAdvisor's funware. *Electronic Markets*, 25, 189–209.

Silva, R. J. R., Rodrigues, R. G., and Leal, C. T. P. (2019). Gamification in management education: a systematic literature review. *BAR-Brazilian Administration Review*, 16, e180103.

Spais, G., Behl, A., Jain, K., Jain, V., and Singh, G. (2022). Promotion and branding from the lens of gamification in challenging times. *Journal of Promotion Management*, 28(4), 413–419.

Tan, F. T. C., Tan, B., Lu, A., and Land, L. (2017). Delivering disruption in an emergent access economy: a case study of an e-hailing platform. *Communications of the Association for Information Systems*, 41(1), 497–516.

Torres, P., Augusto, M., and Neves, C. (2022). Value dimensions of gamification and their influence on brand loyalty and word-of-mouth: relationships and combinations with satisfaction and brand love. *Psychology and Marketing*, 39(1), 59–75.

Wang, L., Gunasti, K., Shankar, R., Pancras, J., and Gopal, R. (2018). Impact of gamification on perceptions of word-of-mouth contributors and actions of word-of-mouth consumers. *Management Information Systems Quarterly*, 44(4), 1987–2011.

Wood, L. C., and Reiners, T. (2015). Gamification. In Encyclopedia of Information Science and Technology, (3rd Edn.). IGI Global, pp. 3039–3047.

Wut, E., Ng, P., Leung, K. S. W., and Lee, D. (2021). Do gamified elements affect young people's use behaviour on consumption-related mobile applications?. *Young Consumers*, 22(3), 368–386.

Xi, N., and Hamari, J. (2019). Does gamification satisfy needs? a study on the relationship between gamification features and intrinsic need satisfaction. *International Journal of Information Management*, 46, 210–221.

Xu, J., Du, H. S., Shen, K. N., and Zhang, D. (2022). How gamification drives consumer citizenship behaviour: the role of perceived gamification affordances. *International Journal of Information Management*, 64, 102477.

Xu, X.-Y., Tayyab, S. M. U., Jia, Q. D., and Wu, K. (2023). Exploring the gamification affordances in online shopping with the heterogeneity examination through REBUS-PLS. *Journal of Theoretical and Applied Electronic Commerce Research*, 18(1), 289–310.

Zhang, L., Shao, Z., Benitez, J., and Zhang, R. (2023). How to improve user engagement and retention in mobile payment: a gamification affordance perspective. *Decision Support Systems*, 168, 113941.

Zhang, L., Shao, Z., Li, X., and Feng, Y. (2021a). Gamification and online impulse buying: the moderating effect of gender and age. *International Journal of Information Management* 61, 113941.

Zhang, Y., Van Horen, F., and Zeelenberg, M. (2021b). Increasing saving intentions through leaderboards: a gamification approach. *PloS one,* 16(4), 249–283.

5 Corporate synergy: a trilateral harmony of financial transparency, information disclosure, and financial performance of Indian corporations

Anupriya Panda[1,a], *Koustubh Kanti Ray*[2,b], *Kumar Mohanty*[2,c] *and Bhuwandeep*[3,d]

[1]Research Scholar, School of Management, KIIT Deemed to be University, Bhubaneswar, India

[2]Professor, School of Management, KIIT Deemed to be University, Bhubaneswar, India

[3]Associate Professor, School of Management, KIIT Deemed to be University, Bhubaneswar, India

Abstract

This empirical study examines the nexus between transparency, information disclosure, and financial performance of Indian entities. Employing a dataset from 2017 to 2022, of Nifty 200 companies, we create a 'financial transparency and information disclosure index' constructed on the Standard & Poor (S&P) benchmark index. The research employs a panel regression model to examine how disclosure and transparency practices affect business profitability, return on equity (ROE) and firm value, (Tobin's Q). Contrary to existing literature, we find an indirect relationship between the Information disclosure index and ROE, suggesting that excessive disclosure may overwhelm investors, leading to higher capital costs and lower ROE. Conversely, a positive impact of the Financial Transparency Index (FTI) on ROE indicates that strong financial transparency enhances investor confidence. Our findings show a positive impact of transparency practices on firm value (Tobin's Q), emphasizing the value of information disclosure in attracting investment and creating shareholder value. This research advances our understanding of the relationships between financial transparency, information disclosure, and firm performance, providing implications for both academia and practitioners.

Keywords: Financial transparency, firm performance, information disclosure, panel regression, S&P benchmark index

Introduction

Stakeholders cannot make prudent financial decisions without relevant financial information (Omran et al., 2008). Concerned stakeholders require information on the enterprise's financial distress condition, as this could cause the failure of the enterprise's profitable operations (Makki and Lodhi, 2014). Inadequate financial disclosures by companies may cause business failures such as the Lehman Brothers collapse, the Enron scandal, the 2008 financial crisis, UTI frauds, the Satyam scam, etc. (Luan and Tang, 2007; Masulis et al., 2007; and Thapar and Sharma, 2017). Businesses strive to keep information as accurate as possible so that unwelcome facts cannot be made public (Shahwan, 2015). Therefore, effective corporate governance requires financial openness and disclosure of information (Mallin, 2016). Transparency and disclosure are highlighted as fundamental practices of corporate governance in the Corporate Governance Principles, (1999) of the Organization for Economic

[a]pandaanupriya@gmail.com, [b]drkkray@gmail.com, [c]kumar@ksom.ac.in, [d]bhuwandeep@ksom.ac.in

DOI: 10.1201/9781003564447-5

Co-operation and Development (OECD). Transparency and disclosure refer to the openness and clarity with which a company communicates its operations and financial performances to its stakeholders, including investors, customers, and the general community (Varshney et al., 2015). Enhanced transparency and effective disclosure practices reduce information asymmetry by eliminating the information disparity between an entity and its stakeholders (Akerlof, 1970). Transparency and disclosures in financial statements give investors a thorough knowledge of a company's financial health, strategic objectives, potential risks, and governance mechanisms. Hence, access to essential company data reduces market volatility (Younas et al., 2021; Bhimavarapu et al., 2022a, b). Companies with a superior level of corporate financial transparency enjoy a lessen capital costs and higher stock prices (Beekes and Brown, 2013). Moreover, transparency and disclosure practices serve as a preventive against the possibility of corporate malpractice (Dewi, 2020).

The Securities and the Exchange Board of India (SEBI) and the Indian Company Act (2013) synergistically play an integral role in fostering transparency, disclosure, and accountability within Indian firms (La Porta et al., 2002). SEBI, as the regulatory authority for the securities market, enforces stringent disclosure requirements for listed companies, ensuring timely and accurate dissemination of financial information (Bujaki and McConomy, 2000). Simultaneously, the Company Act of 2013 provides the legal framework for corporate governance practices, financial reporting standards, board composition, and shareholder rights (Huang and Khanna, 2003). Clause 49 in the Listing Agreement further reinforces SEBI's role in promoting transparency by mandating strict corporate governance practices, including independent director appointments, audit committees, and whistleblower mechanisms (Li and Nair, 2007). The collaborative efforts of SEBI and the Companies Act create a comprehensive system that monitors and regulates market conduct and establishes the necessary legal and governance framework for companies (Chakrabarti and Megginson, 2009).

Existing literature on the relationship between transparency practices and enterprise financial performance yields mixed findings. Hodgson et al. (2011), Black et al. (2006), and Huang and Kung (2010) assert a positive link between CSR disclosure, board separation, and ownership structure, and firm profitability. Conversely, Wahba (2015), Hambrick and D'Aveni (1992), and Daily and Dalton (1994) demonstrate a negative association between corporate transparency, including environmental disclosure, tax transparency, board size, and financial performance. Omran et al. (2008) suggest no significant correlation between disclosure practices and enterprise financial performance. Despite the mixed results, the specific effect of financial transparency, and information disclosure practices on the enterprise' financial performance, particularly in the context of Indian businesses, remains an underexplored research area. Thus, this study aims to explore the nexus between financial performance and the financial transparency and information disclosure practices of Indian firms.

The research article is arranged as: The second section gives a detailed review of past literature with research hypotheses. In the third section, we outline the data and methodology employed in this study, comprising a detailed explanation of the variables, the econometric models employed, and the sample data. Subsequently, in the fourth section, the empirical findings are presented. Finally, the fifth section summarizes the derived conclusions.

Review of Literature and Hypothesis

The past literature recommended several means through which transparency and disclosure practices can affect an enterprise's value and profitability. Previously, studies focused on governance transparency including the depth of corporate governance practices (Habib, 2008) and CSR transparency that reports social and environmental initiatives to promote responsible business conduct (Aksu and Kosedag, 2006). However, contemporary research expanded the concept of transparency to include corporate financial reporting, acquiring and sharing private information, and information broadcasting (Bushman, et al., 2004). The financial reporting of firms covers financial disclosure norms, as per accounting standards (Ugur and Ararat, 2006). Private information acquisition and communication emphasizes the role of financial analysts, institutional investors, and insider trading (Vander Bauwhede and Willekens, 2008). Information dissemination focuses on the impact of media coverage of corporate activities (Millar et al., 2005). Thus, the current study explores corporate transparency from two broad aspects: the firm's financial transparency and the firm's information disclosure. Thus, the study is structured to address three central research questions:

i. How does the overall "financial transparency and information disclosure index" influence firm profitability, measured by return on equity (ROE), and firm value, as measured by Tobin's Q, in Indian corporate entities?
ii. How does the level of financial transparency, measured by the FTI affect firm profitability (ROE) and firm value (Tobin's Q) in Indian corporate entities?
iii. How does the level of information disclosure, measured by the "Information Disclosure Index," impact firm profitability (ROE) and firm value (Tobin's Q) in the Indian corporate sector?

To address these research questions, the study has developed three research objectives:

- To study the influence of the financial transparency and information disclosure index on two vital aspects of enterprise performance: enterprise profitability (assessed through ROE) and enterprise value (assessed through Tobin's Q).
- To evaluate the independent impact of the FTI on firm performance, specifically in terms of firm profitability (ROE) and firm value (Tobin's Q).
- To study the independent impact of the information disclosure index affects firm performance, particularly in relation to firm profitability (ROE) and firm value (Tobin's Q).

The Financial Transparency and the Information Disclosure and Firm Profitability

Several prior works established a direct relationship between an enterprise's transparency and disclosure practices and its ROE. Dye's (1985); and Hermalin and Weisbach (1991) study suggested that the timely, accurate, and comprehensive reporting of financial information enhances ROE by boosting investor confidence and trust. In accordance, Lang and Lundholm (1993); and Shleifer and Vishny (1997) emphasize the significance of financial disclosure quality, revealing a robust positive correlation between high-quality disclosures and ROE. Verrecchia (2001) contends that financial information disclosure mitigates information asymmetry between managers and investors, resulting in increased investor willingness to trade and, consequently, higher profitability. Botosan and Plumlee (2002) and La Porta et al. (2002) suggest that good financial transparency practice diminishes monitoring costs, thereby prompting a reduced rate of return requirement for holding stocks. Bushman et al. (2004) assert that the provision of reliable financial information empowers stakeholders to make well-informed decisions, leading to a more efficient allocation of capital and an overall enhancement of ROE. Eccles and Krzus (2010), emphasize the positive influence of amalgamating financial and non-financial information on transparency and ROE. Uyar and Kilic (2012); and Hope et al. (2013) find that firms with good financial disclosure practices exhibit lower costs of equity compared to those with limited transparency, thereby facilitating more efficient capital allocation and consequent improvement in ROE. Mishra and Bhagwat (2020) explore how transparency and disclosure (T&D) practices influence a firm's vulnerability to financial profitability, particularly considering the competitive landscape. Thus, drawing from the above literature, the present study proposes the following hypothesis:

H1a: There exists a positive correlation between the firm's Financial Transparency and Information Disclosure Index and profitability.

The principles of corporate transparency and comprehensive information disclosure constitute fundamental components of the OECD (1999) and are recognized as pivotal factors influencing the quality of corporate governance. Beaver (2002) advocates for regulatory measures that reduce information asymmetry, maintaining "an equal opportunity" for experienced and inexperienced investors by instituting minimal transparency standards. Lang and Lundholm (1993); and Healy et al. (1999) suggest that executives anticipating equity or debt issuances, willingly disclose crucial information as voluntary disclosure diminishes asymmetric information, thereby reducing the external financing costs for the firm and better ROE. Chi (2009) stated that organizations with more robust corporate governance practices tend to show more informative disclosures, leading to conflict-of-interest problems between executives and owners, ultimately impacting firms' performance. In contrast, Gompers et al. (2003), Lazarides and Pitoska (2009), and Chao, et al. (2010) demonstrate that higher information transparency correlates positively with ROE and thus experiences higher stock returns. Bhimavarapu et al. (2022a, b) further investigate this link by examining the relationship between T&D, financial distress, and the valuation of banks in India. Desai et al. (2008) explore the link between the quality of financial reporting and the cost of equity capital for Indian firms. Their findings suggest that higher-quality financial reporting, which often goes hand-in-hand with strong disclosure practices, can lead to a lower cost of capital for firms. Finally, Dutt and Ghosh (2016) investigate the impact of adopting International Financial Reporting Standards (IFRS) on the quality of financial reporting in India. Their study suggests that implementing these standardized reporting requirements can improve the transparency and overall reliability of financial information for Indian companies. Thus, drawing from the above literature, the current study suggests the following hypothesis:

H1b: The firm's Information Disclosure Index and profitability are positively correlated.

In contrast to earlier research, Chen et al. (2003) claim that excess transparency by businesses increases their operational expenses and faces disclosure risks, which leads to competitors learning confidential information which in turn, results in increased transaction expenses and adverse effects on firm profitability (ROE). Lambert et al. (2007); and Hirshleifer and Teoh (2009) assert that an excess of financial information transparency, if not judiciously managed, can negatively impact a firm's ROE, emphasizing the necessity of maintaining a level of confidentiality to preserve a competitive advantage. The study of Lang et al. (2003); and Core et al. (2006) implies that a firm's private information is valued in the securities market as a risk factor, and an excessive level of information transparency posits an adverse effect on shareholder value creation and, firm profitability, ROE. Chen et al. (2010); and Bushee (2012) contend that increased transparency may attract transient institutions employing aggressive trading strategies and short investment horizons, thereby amplifying a firm's stock return volatility. Li et al. (2020) concluded that financial transparency is associated with diminished ROE due to weakened corporate governance and less mature investor protection mechanisms. Interestingly, Rao et al. (2023) introduces another layer of complexity, investigating how T&D affects a firm's valuation in India while considering the influence of environmental, social, and governance (ESG) factors. Therefore, drawing from the above discussions, the current study proposes the below-mentioned hypothesis:

H1c: The firm's FTI and profitability are negatively correlated.

The Financial Transparency and the Information Disclosure and Firm Value

The disclosure practices of firms exert significant influence over corporate behavior, aiming to safeguard shareholders' interests and enhance firm value. As suggested by Leuz and Wysocki (2008), regulatory environments play a crucial role in shaping companies' approaches to disclosure practices, subsequently impacting how investors interpret disclosed information and, consequently, a firm's overall value. Similarly, Hail, (2008) contends that industry-specific dynamics, including competition levels and technological disruption, introduce complexity into the perception of financial transparency and its subsequent effect on Tobin's Q. Newman and Sansing (1993) assert that corporate information disclosures can unintentionally expose valuable evidence to competitors, diminishing firm value. La Porta et al. (2002) found a negative association between information asymmetry and Tobin's Q. Shin (2003) and Black et al. (2006) link disclosure costs, inefficient governance standards, and negative Tobin's Q. Ball et al. (2012) perspective introduces strategic manipulation of information disclosure, potentially leading to lower Tobin's Q ratios despite high transparency levels. According to Daske and Gebhardt (2006), following international accounting standards emerges as a critical factor influencing how financial information is disclosed and perceived across different countries, contributing to variations in market valuation. In a similar context, Bamber (2011) contended that an exploration of market reactions to earnings announcements reveals insights into the immediate and long-term impacts of transparent disclosure practices on Tobin's Q, shedding light on investor perceptions and valuation dynamics. Furthermore, an examination of the qualitative and quantitative aspects of disclosure practices reveals potential variations in their impact on market perceptions and firm valuation (Botosan and Plumlee, 2002). Also, Lang and Lundholm (1993) stated that the influence of financial analysts as mediators in the dissemination and interpretation of disclosed information is explored, recognizing their role in shaping market reactions to transparency. The interplay of market sentiment, driven by factors such as news and social media, introduces an additional layer of complexity in understanding how transparency practices impact Tobin's Q (Aerts et al., 2007). Distinguishing between long-term and short-term effects, along with assessing the impact of financial crises on the observed relationship, further enriches our comprehension of the multifaceted dynamics between financial transparency, disclosure practices, and Tobin's Q (Hail et al., 2010). Zhou et al. (2018) suggest that excessive financial transparency may lead to lower market valuations, while Hasan, et al. (2014) and Kaewhanam, et al. (2017) identify an adverse association between the firm value (Tobin's Q) and transparency practices and. Akshita and Sharma (2015) find higher transparency to be costly for firms, impacting performance. Taddei and Enriques's (2016) meta-analysis reveals a consistently negative impact of information disclosure on Tobin's Q. Arora et al. (2017) offer insights into the broader impact of disclosure, including CSR disclosure, on firm performance. Thus, the study suggests the following hypotheses based on the above discussion.

H2a: Financial Transparency and Information Disclosure Index exert a negative effect on enterprise value.

H2b: The FTI exerts a negative influence on enterprise value.

H2c: The Information Disclosure Index exerts a negative effect on enterprise value (Tobin's Q).

The Study Design

Data

The data employed for the current empirical study comprises of Nifty 200 companies for 6 years from 2017 to 2022. The sample excludes the banking and financial companies due to their different financial reporting standards, and companies lacking data for more than two years consecutively are also excluded, resulting in a final panel dataset of 145 companies. Financial data, including the ROE, data for Tobin's Q estimation, total assets, leverage ratio, liquidity ratio, and operating profit, is obtained from the Centre for Monitoring Indian Economy (CMIE) database. The parameters for the study's Financial Transparency and Information Disclosure Index are referred from the S&P Transparency and Disclosure Benchmark Index (2002). The parameters of the financial transparency and information disclosure index of the final dataset of 145 companies are assessed from their annual reports from 2017 to 2022.

Variable Measurement and Model Specifications

The Variables' Measurement

Response Variable

The response variable of the study is the enterprise's financial performance. The study employs accounting-based (ROE) as well as market-based metrics (Tobin's Q) as financial performance determinants. ROE, a widely utilized accounting-based metric, focuses on shareholder equity. ROE represents the profitability of a company relative to its shareholder equity, providing a direct measure of how effectively a firm is generating profits from the shareholders' investments. This focus on equity aligns with the study's emphasis on assessing financial performance from a shareholder perspective. Tobin's Q compares the market value of a company's assets to their replacement cost, indicating whether the firm is undervalued or overvalued in the market. This metric accounts for not only the book value of assets but also the market's perception of their worth, thereby providing a clearer evaluation of firm value and investment opportunities.

Explanatory variable

The explanatory variable is the firm's financial transparency and information disclosure index (TDI) scores. The "Financial Transparency and Information Disclosure Index" is constructed using 31 parameters. The development of this index is based on S&P Transparency & Disclosure Benchmark Index (2002). The index is constructed by adding a score of 1 representing the transparency and information disclosure practices followed by the sample firms, whereas 0 denotes otherwise. The financial performance of enterprises is assessed by employing two indicators: ROE, and Tobin's Q. This index is further divided into two indices: (i) FTI scores and (ii) IDI. As a result, the research includes three metrics for measuring the enterprise's transparency: (1) the **Financial**

Transparency and Information Disclosure Index score (TDI); (2) the FTI score, and (3) the Information Disclosure Index Score (IDI). **The FTI is constructed by employing three key dimensions:** (i) the corporate focus, (ii) the accounting policy assessment; and (iii) the accounting policy details. These dimensions collectively encompass 23 distinct parameters. Specifically, the business focus dimension comprises 14 parameters, the accounting policy review dimension encompasses six parameters, and the accounting policy details dimension includes three parameters. The FTI assigns a substantial weightage of 76% (0.76) to these 23 parameters to evaluate and score the firms under consideration. **The TDI is constructed by employing two key dimensions:** (i) the related party transactions; and (ii) the information about auditors. These dimensions collectively encompass eight distinct parameters. Precisely, related party transactions comprise four parameters; and the information about auditors includes four parameters. The IDI assigns a substantial weightage of 24% (0.24) to these eight parameters to evaluate and score the firms under consideration. **The overall index, the Financial Transparency, and Information Disclosure Index (TDI), of 31 parameters is derived by combining the individual scores of the other two indicesthe.** This composite index, which carries a maximum attainable score of 100, serves to assess and compare the effect of the FTI and TDI on the overall evaluation of enterprise performance. The parameters for scoring of the companies are described in Table 5.1 below.

Control Variables

The study further included a group of control variables in the regression model. Previous research work demonstrated that higher levels of leverage increase a company's default risk, increasing capital costs and negatively affecting corporate performance (Jandik and Makhija, 2005). Hence, firm leverage is assessed as the long-term debt divided by the total assets for its possible adverse relation with enterprise performance. The study employs enterprise size, the natural logarithm value of the enterprise's total assets, as larger enterprises tend to be more transparent than smaller firms, and this transparency can lead to improved performance (Rastogi and Kanoujiya, 2021). The study employs the operating profit to analyze the correlation between transparency and disclosure scores and future operating performance (Aerts et al., 2007). The study also employs liquidity ratio to determine the capability of enterprises to meet short-term financial obligations (Chen et al., 2010).

The Model Specifications

The current research article proposes six regression models to verify the association between the enterprise's financial performance and transparency and disclosure index.

$$ROE_{it} = \beta_1 TDI + \beta_2 ES + \beta_3 LV + \beta_4 LQ + \beta_5 OM + \epsilon \quad (1)$$

$$TQ_{it} = \beta_1 TDI + \beta_2 ES + \beta_3 LV + \beta_4 LQ + \beta_5 OM + \epsilon \quad (2)$$

$$ROE_{it} = \beta_1 FTI + \beta_2 ES + \beta_3 LV + \beta_4 LQ + \beta_5 OM + \epsilon \quad (3)$$

$$TQ_{it} = \beta_1 FTI + \beta_2 ES + \beta_3 LV + \beta_4 LQ + \beta_5 OM + \epsilon \quad (4)$$

$$ROE_{it} = \beta_1 IDI + \beta_2 ES + \beta_3 LV + \beta_4 LQ + \beta_5 OM + \epsilon \quad (5)$$

$$TQ_{it} = \beta_1 IDI + \beta_2 ES + \beta_3 LV + \beta_4 LQ + \beta_5 OM + \epsilon \quad (6)$$

where TQ_{it} is Tobin's Q for a company "i "in the year "t," ROE_{it} is the annual ROE for a company "i" in the year "t." TDI is

Table 5.1 Transparency and disclosure index questions applied to Indian companies.

Financial transparency index
Business focus
• Are there any published things concerning corporate strategy?
• Do the enterprises' report specify the nature of the enterprises' affair?
• Do the corporations offer insights into the industry trends?
• What are the details of the products or services provided?
• Is there a breakdown of business segments?
• Does the business expose its market share in any of its sectors?
• Are there any basic earnings forecasts disclosed?
• Is there disclosure of output in physical terms?
• Are there output forecasts provided?
• What are the features of the assets employed by the corporation?
• Are productivity indicators such as ROA and ROE provided?
• Are there any industry-related ratios revealed?
• Does the corporation outline its investment tactics for the forthcoming years?
• Are the details of the firm's investment plans for the future disclosed?
Accounting policy review
• Do the corporations offer quarterly financial data?
• Is there an analysis of the business's accounting policy?
• Does the firm reveal the accounting standards it adheres to?
• Are accounts provided according to local accounting standards?
• Are accounts presented using internationally recognized accounting methods?
• Are balance sheets, income statements, and cash-flow statements provided using universally recognized approaches?
Accounting policy details
• Are approaches of asset valuation revealed by the business?
• Is information provided concerning the method of depreciation for fixed assets?
• Does the corporation produce consolidated financial statements?
Information disclosure index
Related party structure and transactions
• Does the corporation list associates in which it holds a minority stake?
• Is the ownership structure of associates unveiled by the business?
• Is there a record of related party transactions?
• Is there a record of group transactions provided?
Information on auditors
• Is the auditing firm's name unveiled by the business?
• Does the corporation comprise the auditors' report?
• Is the amount paid in audit payments to the auditor revealed?
• Are any non-audit payments salaried to the auditor disclosed by the business?

Source: Patel, Sandeep A., and George S. Dallas' S&P T&D Benchmark Index (2002)

Table 5.2 Variables used in the regression model.

Variable	Nature	Symbol	Definition/Estimation method	Source
Return on equity	Dependent	ROE	Ratio of net income/shareholder equity. To measure the firm profitability	CMIE prowess IQ
The Tobin's Q value	Dependent	TQ	The ratio of the enterprise's total market value/total assets value. To measure firm value	CMIE Prowess IQ
Financial transparency and information disclosure index	Independent	TDI Score	Depicts the financial transparency and information disclosure index score of firms. An index is developed comprising 31 parameters	Annual Reports
FTI	Independent	FTI score	It represents the FTI score of firms and is computed by developing an Index encompassing 23 parameters	Annual Reports
Information disclosure index	Independent	IDI score	It represents the information disclosure index score of firms and is computed by developing an Index encompassing eight parameters	Annual Reports
Enterprise size	Control	ES	Enterprise size is the natural logarithm of firm total assets. To measure the impact of the size of the firm on transparency and disclosure levels.	CMIE Prowess IQ
Leverage ratio	Control	LV	The leverage ratio is the long-term debt/ total asset. To measure the default risk of firms.	CMIE Prowess IQ
Liquidity ratio	Control	LQ	Liquidity refers to the ratio of current assets to current liabilities, a measure of a company's capability to fulfil immediate financial responsibilities.	CMIE Prowess IQ
Operating profit margin	Control	OM	EBITDA, which stands for earnings before interest, taxes, depreciation, and amortization, serves as an indicator of a company's operational performance.	CMIE Prowess IQ

Source: Author's compilation

the enterprise's transparency and disclosure index score. FTI is the FTI score and IDI is the information disclosure index score. Enterprise size is the natural logarithm of total assets; leverage is the ratio of long-term debt to total assets; liquidity is the ratio of current assets to current liabilities; and operating margin is the earnings before interest, tax, depreciation, and amortization (EBITDA). The variables' details and definitions are provided in Table 5.2.

Results and Interpretations of the Study

Summary Statistics

Table 5.3 presents the descriptive statistics for six years from 2017 to 2022, drawing from a dataset of 870 observations. The average value of ROE is 0.0689, indicating moderate profitability. The average Tobin's Q is 3.696, signifying that, on average, companies are valued at a premium relative to their book value. The mean value of

Table 5.3 Summary statistics.

Variables	Observations	Average	Standard deviation	Min. value	Max. value
ROE	870	.0688501	2.824852	-82.21393	4.672803
TQ	870	3.695818	4.048127	.0084629	27.60404
TDI	870	81.66192	4.211504	67.56522	90.08696
FTI	870	81.22939	4.316295	65.21739	86.95652
IDI	870	83.03161	11.08851	62.5	100
Enterprise size (ES)	870	11.7982	1.427979	-.9162907	16.08982
Leverage (LV)	870	.1797515	.2678591	.0016576	5.280621
Liquidity (LQ)	870	1.846115	1.5368	.0496492	20.97222
Operating margin (OM)	870	39.47854	299.999	–438.9209	7851.428

Note: A strongly balanced panel data set with 870 total observations
Source: Computed by author

Table 5.4 Correlation and variation inflation factor (VIF).

Variables	T&D index	FTI	IDI	Firm size	Leverage	Liquidity	Operating margin	VIF
TDI	1							1.01
FTI	0.7751	1						1.01
IDI	0.6272	0.0061	1					1.00
Enterprise size	0.0293	0.0491	0.0142	1				1.06
Leverage	–0.0019	0.0297	-0.0396	0.1344	1			1.08
Liquidity	0.0693	0.0719	0.0210	0.2225	0.2526	1		1.13
Operating margin	0.0296	0.0481	–0.0125	0.0232	–0.0301	0.0848	1	1.01
								1.04

Note: Variance inflation test (VIF) is conducted for the multicollinearity issues
Source: Estimated by the author

the TDI is 81.662, indicating a high level of transparency among the sampled firms. The average FTI is 81.229, reflecting a strong commitment to financial transparency. The average IDI is 83.032, suggesting a relatively high level of information disclosure. The average enterprise size is 11.798, signifying a moderately sized sample. With an average leverage ratio of 0.180, the sampled firms maintain relatively low financial leverage on average. The average liquidity ratio is 1.846, suggesting reasonable liquidity levels. The average operating profit margin is 39.48%, signifying profitability in core operations.

Correlation and Multicollinearity Test

Table 5.4 summarizes the correlation matrix and variance inflation factor (VIF) test results. Key findings include strong positive correlations between the FTI and TDI, as well as between the TDI and the IDI.

Weak positive correlations are observed between TDI and enterprise size, liquidity, and operating margin. Conversely, there is a very weak negative correlation between TDI and Leverage. To measure the multicollinearity problem, the Variance Inflation Factor is used. A VIF value above 5 is considered problematic. The VIF is found to be 1.04, which proposes that multicollinearity is not a significant issue in the study.

Regression Results

Model (A) firm profitability (ROE) as the dependent variable: TDI exhibits a direct association with ROE, indicating that a single-unit rise in the TDI is associated with an estimated ROE increase of approximately 0.0367 units. Similarly, FTI and IDI also positively impact ROE, with one-unit increases corresponding to estimated ROE increases of around 0.0355 and 0.0050 units, respectively. Conversely, Enterprise Size positively impacts ROE, while Leverage exerts a negative influence, implying that larger firms tend to have higher ROE, while higher leverage tends to decrease ROE. Liquidity positively affects ROE, and Operating Margin posits a minor negative effect.

Model (B) enterprise value (Tobin's Q) as the response variable: TDI with a positive coefficient value of approximately 0.0007, indicates that an increase in the value of TDI is related to an estimated rise of approximately 0.00073 units in Tobin's Q. This suggests that enhanced transparency and disclosure practices can positively contribute to a company's market valuation. FTI demonstrates a positive effect on enterprise value, with a coefficient value of approximately 0.019, implying that factors of FTI are linked to higher market valuations. Similarly, IDI also exerts a positive impact, with a coefficient value of approximately 0.0095291. Firm Size emerges as a significant contributor to Tobin's Q, with a notable coefficient value of approximately 1.131, suggesting that larger firms are allied with higher market valuations, likely due to their market dominance and potential for future growth. Conversely, leverage has a negative influence on Tobin's Q, as indicated by its coefficient value of approximately –0.3505. This suggests that higher leverage levels may be viewed unfavorably by the market in terms of firm value. Additionally, liquidity exerts a positive effect on enterprise value, underscoring the importance of having sufficient liquid assets, while operating margin positively influences Tobin's Q, implying that higher operating margins are linked to greater market valuation.

Panel Regression Outcome

The panel data regression analysis for the panel dataset containing 145 Indian enterprises is summarized in Table 5.6. Pooled ordinary least square (OLS) and suitable fixed effect model (FE) or random effect model (RE) are part of the panel data analysis. The Breusch-Pagan LM test is used to determine whether the data are heteroskedastic before conducting an empirical data analysis. Heteroskedasticity was indicated by the test findings (χ^2 = 263.20 and p = .000). Table 5.5 summarizes the findings of the simple regression (OLS). The calculated OLS results are heterogeneity biased and the firms' attributes are only partially explained by the explanatory variables of the OLS model. Thus, OLS is not a trustworthy tool for panel data analysis because of the biased and inconsistent estimate of the explanatory variable's coefficient. To account for the heterogeneity, panel analysis is performed using regression models 1 through 6 and the balanced panel dataset. The Hausman test is used to determine whether fixed effect (FE) or random effect (RE) models are appropriate for panel regression. The FE and RE models' estimated coefficients are compared using the Hausman test (Hausman, 1978). The RE model's effectiveness and consistency serve as the test's null hypothesis. As opposed to

Table 5.5 Regression outcomes.

	Response variable (firm performance)			
	Return on equity (ROE)		Tobin's Q (TQ)	
Independent variable	Coefficient values	Standard error	Coefficient values	Standard error
TDI	.0367307 (0.108)	.0228138	.0007353**(0.028)	.0298538
FTI	.0355832***(0.011)	.0222894	.0190623** (0.013)	.0291596
IDI	.0050291** (0.062)	.00866	.0095291***(0.004)	.0113129
Enterprise size	.0832988*(0.029)	.0691871	1.131442***(0.000)	.0904183
Leverage	-.4204406*(0.058)	.3716523	-.3505208 (0.070)	.4851149
Liquidity	.0024086*(0.041)	.0660559	.1292132** (0.036)	.0864934
Operating margin	-.0001655*(0.056)	.0003211	.0009533*(0.023)	.0004195

The "p" values are in parentheses.

**p < .05, *** p < .001, and *p < .10

Source: Author's estimation

Table 5.6 Panel regression results (FE model).

	Dependent variable (firm performance)			
	ROE		Tobin's Q	
Independent variable	Coef. values	Std. error	Coef. values	Std. error
TDI	.0310361 (0.225)	.0272446	.0177696***(0.004)	.0186089
FTI	.0332158***(0.022)	.0302119	.0464775** (0.024)	.0205746
IDI	.0048478 (0.093)	.0090734	.0050575***(0.014)	.0061941
Enterprise size	.2951034**(0.040)	.219842	.250284** (0.034)	.1500787
Leverage	–.3194237 *(0.051)	.6322112	3.690136***(0.000)	.4321572
Liquidity	.0638254 (0.082)	.1159078	.0276399** (0.027)	.0791263
Operating margin	.0000564 (0.067)	.0004491	.0008348* (0.007)	.0003066
R^2	0.527		0.572	

The "p" values are in parentheses.

**p < .05, ***p < .001, and *p < .10

Source: Author's estimation

this, the alternate hypothesis assumed that the FE regression model would be reliable and operative. The Husman test's outcome ($\chi^2 = 65.50$ and the $p = .00$) indicates the FE model is appropriate for additional research since the null hypothesis is rejected. The regression model 1 examined the impact of the TDI on ROE. Whereas regression model 2 assessed the effect of the TDI on enterprise value. Likewise, regression model 3 analyses the effect of the FTI on the firm's ROE, and the contrary model 4 tests the effect of the FTI on enterprise value. Moreover, regression model 5 analyzes the influence of the

IDI on the ROE of firms, and in contrast, model 6 tested the impact of IDI on Tobin's q, treating the control variables in all models as constants. The R-square value for firm profitability (ROE) and the firm value (Tobin's q) are observed to be 52.7% and 57.2%, respectively. These results indicate that changes in the response variable i.e., firm performance, are adequately explained by the explanatory variables.

In Model (A) with firm profitability (ROE) as the dependent variable, the FTI exhibits a highly positive influence on ROE, whereas leverage has a highly negative impact. Additionally, firm size is positively associated with ROE, and liquidity shows a marginally significant positive relationship. However, TDI, IDI, and operating margin do not significantly affect ROE. Consequently, the study rejects Hypothesis (H1a) and concludes that the TDI possesses an adverse effect on firm profitability (ROE). Similarly, hypothesis (H1b) is also rejected, concluding that the IDI negatively influences firm profitability (ROE). Additionally, the rejection of hypothesis (H1c) suggests that the FTI has a detrimental impact on firm profitability (ROE). In model (B) with firm value (Tobin's Q) as the dependent variable, the TDI, FTI, IDI, firm size, leverage, and liquidity demonstrate significant positive associations with Tobin's Q, indicative of their influence on market valuation. Although the operating margin exhibits marginal significance, it also positively affects Tobin's Q. Therefore, the study rejects hypothesis (H2a), concluding that the TDI exerts a positive effect on enterprise value (Tobin's Q). Similarly, the rejection of Hypothesis (H2b) proposes that the FTI positively impacts Tobin's Q. Furthermore, the rejection of hypothesis (H2c) indicates that the IDI positively influences Tobin's Q.

Robustness Test

In conducting robustness tests, the study employs different variables such as return on assets (ROA) and price-to-book ratio (PBR), distinct from the initial test's focus on ROE and Tobin's Q. This approach allows for a comprehensive examination of the stability and generalizability of the main findings by assessing the impact of alternative metrics on the results. By incorporating ROA and PBR in the robustness tests, the study evaluates the sensitivity of conclusions to variations in financial performance indicators, providing insights into the robustness of the initial findings and enhancing the reliability of the research outcomes. Table 5.7 presents the outcomes of the robustness test conducted to validate the initial results. For the dependent variable, ROA, the TDI, FTI, IDI, firm size, and liquidity all show statistically significant coefficients. Specifically, FTI and Firm Size have positive coefficients, indicating a positive impact on ROA, while leverage has a negative coefficient, suggesting a negative impact. The TDI and IDI also demonstrate positive effects on ROA. This suggests that higher levels of transparency and financial disclosure are associated with improved ROA. For the dependent variable PB ratio, similar trends are observed. FTI, firm size, liquidity, and leverage have statistically significant coefficients. FTI and liquidity exhibit positive coefficients, indicating a positive impact on PB ratio, while leverage has a negative coefficient, implying a negative effect. This suggests that higher financial transparency, larger firm size, and greater liquidity are associated with higher PB ratio, while higher leverage is associated with lower PB ratio. Overall, this robustness check provides support for the initial findings of the research paper, indicating that transparency-related factors, firm size, liquidity, and leverage play significant roles in explaining variations in firm performance, whether measured by ROA or PB ratio. The reported coefficients and their significance levels help strengthen the paper's conclusions and the reliability of the relationships identified in the study. The R-square value for firm profitability (ROE)

Table 5.7 Robustness check result.

	Dependent variable (Firm performance)			
	ROA		PB ratio	
Independent variable	Coef. values	Std. error	Coef. values	Std. error
TDI	.0256206 (0.107)	.0116247	.0113532***(0.038)	.0248979
FTI	.0446713***(0.021)	.0112384	.0341869***(0.037)	.0540848
IDI	.0141382 (0.057)	.00766	.0241048***(0.029)	.0943951
Enterprise size	.0723798***(0.042)	.0782652	1.34377***(0.035)	.0754150
Leverage	-.2313205 (0.62)	.2525412	-4.15048***(0.016)	.0451067
Liquidity	.0113154***(0.036)	.0551324	.0191321***(0.025)	.0207193
Operating margin	–.0121321 (0.055)	.0002322	. .0004534**(0.046)	.0035002
R^2	0.517		0.552	

The "p" values are in parentheses.

**p < .05, ***p < .001, and *p < .10

Source: Author's estimation

and the firm value (Tobin's q) are observed to be 51.7% and 55.2%, respectively. These results indicate that changes in the response variable i.e., firm performance, are adequately explained by the explanatory variables. Thus, the robustness of the model is addressed with pre- and post-estimation diagnostics techniques.

Discussion

Transparency and disclosure play a fundamental role in shaping corporate performance. This study investigates how the TDI impacts firm performance, ROE and Tobin's Q. The study employs the TDI, FTI, and IDI as explanatory variables. The findings reveal that both the TDI and IDI have no significant effect on ROE. Instead, the study suggests an inverse relationship, indicating that excessive information disclosure resulting from high transparency levels may overwhelm investors, leading to higher capital costs and lower ROE. This contrasts with prior research by Aggarwal and Goodell (2013). Contrarily, the study discovered a positive effect of FTI, on ROE, supporting the previous research by Durnev and Kim (2005). This highlights that firms with strong financial transparency in their reporting enhance investor confidence, leading to higher stock prices and increased ROE. Additionally, the paper revealed a positive association between the TDI and enterprise value, aligning with an earlier study by Zingales and Rajan (2003). This underscores the value of transparency in attracting investment and creating shareholder value. Transparent firms may disclose negative information, impacting ROE, but enhance their perceived ability to create future value, reflected in higher Tobin's Q.

The study's findings suggest that while transparency, as measured by the transparency and disclosure index (TDI), plays a crucial role in attracting investment and creating shareholder value, excessive information disclosure may have adverse effects on firm performance, particularly on ROE. Firms should therefore strive to strike a balance between transparency and avoiding information overload, ensuring that disclosures are relevant and easily interpretable for investors. Moreover, the positive impact of the FTI on ROE underscores the importance of clear and comprehensive financial reporting in building investor confidence and driving stock prices. Managers should prioritize enhancing financial transparency practices to align with investor expectations and preferences, ultimately driving higher ROE and Tobin's Q while avoiding the pitfalls of excessive information disclosure.

Conclusion

This empirical research enhances our understanding of the complex relationship between transparency and disclosure practices and the financial performance of Indian companies. Particularly, it broadens the existing knowledge by investigating this relationship within a diverse sample of Nifty 145 companies for six years from 2017 to 2022. The empirical results state that the return of investment (ROE) is negatively impacted by TDI and the IDI, but the FTI exhibits a positive correlation. This underscores the importance of striking an optimal balance between financial transparency and disclosure practices to maximize profitability. Furthermore, the study highlights that all three transparency and disclosure indices have a direct influence on firm value, (Tobin's Q), emphasizing their pivotal role in enhancing market valuation. This underscores the critical imperative for firms to meticulously assess their transparency and disclosure strategies. They need to align these strategies with their strategic objectives and meet the expectations of a diverse array of stakeholders. This necessitates a careful balancing act between transparency and the exigencies of confidentiality in certain areas of operation.

However, it is vital to recognize the limitations of this study. The study only relies on quantitative data and overlooks the qualitative aspects that can significantly impact transparency and disclosure practices. Additionally, its focus only the on Indian firms may restrict the findings to more diverse contexts. These limitations offer avenues for improvement in future research work. Succeeding studies may consider incorporating qualitative data to provide a better understanding of the dynamics that govern transparency and disclosure practices. Expanding the scope to encompass firms from diverse geographic regions to provide a more inclusive perspective. Furthermore, a more thorough examination of external contextual factors could be integrated to offer a more thorough comprehension of the complex connection between financial performance, disclosure, and transparency.

References

Aerts, W., Cormier, D., and Magnan, M. (2007). The association between web-based corporate performance disclosure and financial analyst behavior under different governance regimes. *Corporate Governance: An International Review*, 15, 1301–1329.

Akerlof, G. A. (1970). The market for 'lemons': quality uncertainty and the market mechanism. *Quarterly Journal of Economics*, 90, 629–650.

Akshita, A., and Sharma, C. (2015). Impact of firm performance on board characteristics: empirical evidence from India. *IIM Kozhikode Society and Management Review*, 4(1), 53–70.

Aksu, M., and Kosedag, A. (2006). Transparency and disclosure scores and their determinants in the Istanbul stock exchange. *Corporate Governance: An International Review*, 14, 277–296.

Arora, A., Jain, P., and Kaur, N. (2017). The impact of corporate governance on transparency and disclosure practices of listed companies in India. *International Journal of Law and Management*, 59(1), 123–132.

Ball, R., Jayaraman, S., and Shivakumar, L. (2012). The effect of mark-to-market accounting for financial assets and liabilities on financial reporting transparency and information asymmetry in banks. *Social Science Research Network (SSRN)*, 6(1), 111–131.

Bamber, M. A. (2011). An examination and assessment of mandatory financial instruments disclosures. PhD diss.

Beaver, W. H. (2002). Perspectives on recent capital market research. *The Accounting Review*, 77(2), 453–474.

Beekes, W., and Brown, P. (2013). Do better-governed Australian firms make more informative disclosures? In Financial Accounting and Equity Markets. Routledge, pp. 233–261.

Bhimavarapu, S. R., Kishore, P. V. V., and Bhagat, M. S. (2022). Impact of transparency and disclosure (T&D) and financial distress (FD) on the valuation of Banks in India. *Journal of Risk and Financial Management*, 15(11), 205.

Black, B. S., Jang, H., and Kim, W. (2006). Does corporate governance predict firms' market values? *Journal of Finance*, 61, 655–687.

Bhimavarapu, V. M., Rastogi, S., and Kanoujiya, J. (2022). Ownership concentration and its influence on transparency and disclosures of banks in India. *Corporate Governance: The international journal of business in society, 22(1), 18–43.*

Botosan, C. A., and Plumlee, M. A. (2002). A re-examination of disclosure level and the expected cost of equity capital. *Journal of Accounting Research*, 40(1), 21–40.

Bujaki, M., and McConomy, B. J. (2002). Corporate governance: factors influencing voluntary disclosure by publicly traded Canadian firms. *Canadian Accounting Perspectives*, 1(2), 105–139.

Bushee, B. J. (2012). Discussion of 'financial reporting opacity and informed trading by international institutional investors.' *Journal of Accounting and Economics*, 54(2-3), 221–228.

Bushman, R. M., Piotroski, J. D., and Smith, A. J. (2004). What determines corporate transparency? *Journal of Accounting Research*, 42, 207–252.

Chakrabarti, R., and Megginson, W. I. (2009). Corporate governance in India. In Global Corporate Governance. Columbia University Press, pp. 151–176.

Chao, C.-F., Hsu, C. C., and Yeh, H. S. (2010). The relationship between information transparency and firm value: evidence from Taiwan. *International Journal of Business Excellence*, 3(2), 125–141.

Chen, K. C., Wei, K. C. J., and Chen, Z. (2003). Disclosure, corporate governance, and the cost of equity capital: evidence from Asia's emerging markets. *Corporate Governance: An International Review*, 11(1), 65–84.

Chen, W.-P., Chung, H. L., Hsu, T. L., and Wu, S. T. (2010). External financing needs, corporate governance, and firm value. *Corporate Governance: An International Review*, 18(3), 234–249.

Chi, L.-C. (2009). Do transparency and disclosure predict firm performance? evidence from the Taiwan market. *Expert Systems with Applications*, 36(8), 11198–11203.

Core, J. E., Guay, W. R., and Larcker, D. F. (2006). Does weak governance cause weak stock returns? an examination of firm operating performance and investors' expectations. *Journal of Finance*, 61, 655–687.

Daily, C. M., and Dalton, D. R. (1994). Corporate governance and the bankrupt firm: an empirical assessment. *Strategic Management Journal*, 15, 643–654.

Daske, H., and Gebhardt, G. (2006). International financial reporting standards and experts' perceptions of disclosure quality. *Abacus*, 42(3-4), 461–498.

Desai, V., Rankin, M., and Venkatachalam, M. (2008). Financial reporting quality and cost of equity capital in India. *The Accounting Review*, 83(3), 861–889.

Dewi, R. R. (2020). Building reputation through environmental disclosure. *Indonesian Management and Accounting Research*, 18, 1–16.

Durnev, A., and Kim, E. H. (2005). To steal or not to steal: firm attributes, legal environment, and valuation. *The Journal of Finance*, 60(3), 1461–1493.

Dutt, S., and Ghosh, S. (2016). The impact of mandatory IFRS adoption on financial reporting quality in India. *Abacus*, 52(2), 221–243.

Dye, R. A. (1985). Disclosure of non-proprietary information. *Journal of Accounting Research*, 23, 123–145.

Eccles, R. G., and Krzus, M. P. (2010). One report: integrated reporting for a sustainable Strategy. John Wiley & Sons.

Gompers, P. A., Ishii, J. L., and Metrick, A. (2003). Corporate governance and equity prices. *Quarterly Journal of Economics*, 118, 107–155.

Habib, A. (2008). Corporate transparency, financial development and the allocation of capital: empirical evidence. *Abacus*, 44(1), 1–21.

Hail, L., Leuz, C., and Wysocki, P. D. (2008). The role of liquidity in the relationship between corporate transparency and risk. Working paper.

Hambrick, D. C., and D'Aveni, R. A. (1992). Top team deterioration as part of the downward spiral of large corporate bankruptcies. *Management Science*, 38, 1445–1466.

Hasan, M. B., Ahsan, A. F. M. M., Rahaman, M. A., and Alam, M. N. (2014). Influence of capital structure on firm performance: evidence from Bangladesh. *International Journal of Business and Management*, 9, 184–194.

Hausman, Jerry A. (1978). "Specification tests in econometrics." Econometrica: Journal of the econometric society: 1251–1271.

Healy, P. M., Hutton, A. P., and Palepu, K. G. (1999). Stock performance and intermediation changes surrounding sustained increases in disclosure. *Contemporary Accounting Research*, 16(3), 485–520.

Hermalin, B. E., and Weisbach, M. S. (1991). The effects of board composition and direct incentives on firm performance. *Financial Management*, 20(4), 101–112.

Hirshleifer, D., and Teoh, S. H. (2009). Thought and behavior contagion in capital markets. In Hens, T., and Schenk-Hoppe, K. R. (Eds.), Handbook of Financial Markets: Dynamics and Evolution. North-Holland, pp. 1–56.

Hodgson, A., Lhaopadchan, S., and Buakes, S. (2011). How informative is the Thai corporate governance index? a financial approach. *International Journal of Accounting and Information Management*, 19, 53–79.

Hope, O.-K., Ma, M. S., and Thomas, W. B. (2013). Tax avoidance and geographic earnings disclosure. *Journal of Accounting and Economics*, 56(2-3), 170–189.

Huang, C.-L., and Kung, F. H. (2010). Drivers of environmental disclosure and stakeholder expectation: evidence from Taiwan. *Journal of Business Ethics*, 96, 435–451.

Huang, Y., and Khanna, T. (2003). Can India take over China? *Foreign Policy*, 30(3), 74–87.

Jandik, T., and Makhija, A. K. (2005). Debt, debt structure, and corporate performance after unsuccessful takeovers: evidence from targets that remain independent. *Journal of Corporate Finance*, 11, 882–914.

Kaewhanam, P., Jiraporn, P., and Siriboon, S. (2017), The relationship between information transparency and firm value: evidence from Thailand. *Journal of Accounting and Finance*, 47(3), 1–21.

La Porta, R., Lopez-de-Silanes, F., Shleifer, A., and Vishny, R. W. (2002). Investor protection and corporate valuation. *Journal of Finance*, 57, 1147–1170.

Lambert, R., Leuz, C., and Verrecchia, R. E. (2007). Accounting information, disclosure, and the cost of capital. *Journal of Accounting Research*, 45(2), 385–420.

Lang, M. H., Lins, K. V., and Miller, D. P. (2003). ADRs, analysts, and accuracy: does cross-listing in the United States improve a firm's information environment and increase market value? *Journal of Accounting Research*, 41(2), 317–345.

Lang, M., and Lundholm, R. (1993). Cross-sectional determinants of analyst ratings of corporate disclosures. *Journal of Accounting Research*, 31(2), 246–271.

Lazarides, T. G., and Pitoska, E. (2009). Disclosure factors of executive managers remuneration: a probit model. *SSRN*. 6(1), 5–28.

Leuz, C., and Wysocki, P. D. (2008). Economic consequences of financial reporting and disclosure regulation: a review and suggestions for future research. *SSRN*, 94(3), 461–88.

Li, C., Liu, X., and Zhang, M. (2020). Financial transparency and disclosure, institutional quality, and firm performance in emerging markets. *International Review of Economics and Finance*, 71, 18–36.

Li, S., and Nair, A. (2007). A comparative study of economic reforms in China and India: what can we learn? *Global Economic Review*, 36(2), 147–166.

Luan, C.-J., and Tang, M. J. (2007). Where is independent director efficacy?" *Corporate Governance: An International Review*, 15, 636–643.

Makki, M. A. M., and Lodhi, S. A. (2014). Impact of corporate governance on intellectual capital efficiency and financial performance. *Pakistan Journal of Commerce and Social Sciences (PJCSS)*, 8, 305–330.

Mallin, C. A. (2016). Corporate Governance. Oxford: Oxford University Press.

Masulis, R. W., Wang, C., and Xie, F. (2007). Corporate governance and acquirer returns. *Journal of Finance*, 62, 1851–1889.

Millar, C. C. J. M., Eldomiaty, T. I., Choi, C. J., and Hilton, B. (2005). Corporate governance and institutional transparency in emerging markets. *Journal of Business Ethics*, 59, 163–174.

Mishra, D. K., and Bhagwat, A. S. (2020). Transparency and disclosure and financial distress of non-financial firms in India under competition: investors' perspective. *Sustainability*, 16(4), 2174.

Newman, P., and Sansing, R. (1993). Disclosure policies with multiple users. *Journal of Accounting Research*, 31, 92–112.

OECD (1999). OECD Principles of Corporate Governance Meeting of the OECD Council at Ministerial Level. http://www.copyright.com/.

Omran, M. M., Bolbol, A., and Fatheldin, A. (2008). Corporate governance and firm performance in Arab equity markets: does ownership concentration matter? *International Review of Law and Economics,* 28, 32–45.

Rao, P. S., Faff, R. W., and Bhaumik, S. K. (2023). Does the impact of transparency and disclosure on the firm's valuation depend on the ESG? *Sustainability,* 15(3), 1822.

Rastogi, S., and Kanoujiya, J. (2022). Does transparency and disclosure (T&D) improve the performance of Banks in India? *International Journal of Productivity and Performance Management.*

Shahwan, T. M. (2015). The effects of corporate governance on financial performance and financial distress: evidence from Egypt. *Corporate Governance*, 15, 641–662.

Shin, H. S. (2003). Disclosures and asset returns. *Econometrica*, 71, 105–133.

Shleifer, A., and Vishny, R. W. (1997). A survey of corporate governance. *The Journal of Finance*, 52(2), 737–783.

Taddei, M., and Enriques, L. (2016). The relationship between transparency and Tobin's Q: a meta-analysis. *Journal of Accounting and Economics*, 61(2-3), 319–344.

Thapar, M., and Sharma, A. (2017). Corporate governance in India: an analysis. *Journal of Economic and Social Development*, 4, 81–87.

Patel, Sandeep A., and George S. Dallas. "Transparency and disclosure: Overview of methodology and study results-United States." Available at SSRN 422800 (2002). Transparency and Disclosure: Overview of Methodology and Study Results—United States.

Ugur, M., and Ararat, M. (2006). Does macroeconomic performance affect corporate governance? evidence from Turkey. *Corporate Governance: An International Review*, 14(4), 325–348.

Uyar, A., and Kilic, B. (2012). The impact of transparency and disclosure on cost of equity: evidence from Turkey. *Journal of Business Finance and Accounting*, 39(3-4), 476–502.

Vander Bauwhede, H., and Willekens, M. (2008). Disclosure on corporate governance in the European union. *Corporate Governance: An International Review*, 16(2), 101–115.

Varshney, P., Kaul, V. K., and Vasal, V. K. (2015). Corporate governance index and firm performance: empirical evidence from India. *LBS Journal of Management and Research*, 13, 59–75.

Verrecchia, R. E. (2001). Essays on disclosure. *Journal of Accounting and Economics*, 32(1-3), 97–180.

Wahba, H. (2015). The joint effect of board characteristics on financial performance: empirical evidence from Egypt. *Review of Accounting and Finance*, 14, 20–40.

Younas, N., Ud Din, S., Awan, T., and Khan, M. Y. (2021). Corporate governance and financial distress: asian emerging market perspective. *Corporate Governance: The International Journal of Business in Society*, 21(4), 702–715.

Zhou, Z., Zhou, H., Peng, D., Chen, X., and Li, S. (2018). Carbon disclosure, financial transparency, and agency cost: evidence from chinese manufacturing listed companies. *Emerging Markets Finance and Trade*, 54(12), 2669–2686.

Zingales, L., and Rajan, R. (2003). *Banks and markets: the changing character of European finance,* 12, 341–364.

6 Cognitive, affective, and demographic factors and their impact on human behavior to adopt LLM-based generative artificial intelligence

Kharabela Rout[1,a], *Priti Ranjan Sahoo*[2,b], *Pratyuesha Behera*[1,c] *and Archana Bhuyan*[1,d]

[1]Research Scholar, School of Management, KIIT Deemed to be University, Bhubaneswar, India

[2]Associate Professor, School of Management, KIIT Deemed to be University, Bhubaneswar, India

Abstract

This research aims to understand the impact of cognitive and affective factors on the adoption of generative artificial intelligence platforms. The proposed conceptual model, which examines cognitive antecedents such as perceived usefulness, perceived ease of use, and relative advantage, and affective antecedents such as pleasure and arousal, with functional value as a key precursor to adoption intention, provides a practical framework for designing and implementing generative artificial intelligence platforms. The study's findings, based on data from 107 respondents in India, including students and professionals, collected through a self-administered questionnaire, reveal the significant positive impact of cognitive factors like relative advantage and functional value on adoption intention. The study also identifies partial and competitive mediation between perceived ease of use, functional value, and intention to adopt generative artificial intelligence contributing to a deeper understanding of adoption behavior with significant practical implications.

Keywords: Affective, arousal, cognitive, generative artificial intelligence, large language model, pleasure

Introduction

In the recent advancements in the field of generative artificial intelligence (AI), the years 2022 and 2023 can be considered landmark years for the progression of generative AI. Generative AI, in general, is a machine learning system trained with a large quantity of data to generate output based on the user's command or prompts (Sætra, 2023). Various models are employed in generative AI, each with billions of parameters designed for diverse tasks. For example, the large language model (LLM), trained on massive text data, is utilized for natural language processing (NLP), aiding the machine learning system in understanding and processing text to generate nearly natural human-like language (Kasneci et al., 2023). OpenAI's Chat generative pre-trained transformer (Chat GPT), which is the acronym for Chat GPT, is an example of an LLM-based generative AI focused on text (Kasneci et al., 2023). Generative AI streamlines tasks across sales, marketing, customer service, and financial reporting while enhancing e-mail correspondence (Fui-Hoon Nah et al., 2023). In education, it facilitates student research, addresses queries, and improves writing skills while assisting teachers in lesson planning and content creation (Chiu, 2023).

[a]rkharabela@gmail.com, [b]prsahoo@ksom.ac.in, [c]pratyuesha.behera1@gmail.com, [d]archana.bhuyan1996@gmail.com

DOI: 10.1201/9781003564447-6

While numerous studies on generative AI adoption have been conducted based on the technology acceptance model (TAM) (Baek and Kim, 2023; Gude, 2023; Ma and Huo, 2023; Tiwari et al., 2023) the exploration of crucial factors in the context of the technology adoption research paradigm, such as cognitive factors like relative advantage (RA) and affective factors like pleasure (PL) and arousal (AUR), influencing the adoption behavior of LLM based generative AI remains unaddressed in the scholarly literature. Functional value (FV), a pivotal factor in technology usage, has received limited attention in studies examining its impact on adoption intention. Furthermore, there is a lack of knowledge on how perceived ease of use (PEoU) and perceived usefulness (PU) impact the FV of generative AI platforms. India, with its growing digital infrastructure and youthful population, emerges as a vital market for generative AI firms despite limited existing research on the Indian context.

Hence, noting the above gaps in the existing body of knowledge on adoption behavior toward generative AI, this study aims to answer two central research questions in the context of India as proposed: How do users use generative AI platforms for their professional activities? What is the demographic, cognitive, and affective factors influence generative AI adoption? This study seeks to construct a conceptual model by amalgamating the TAM, Consumer Acceptance of Technology (CAT), and theory of value consumption (TVC). It aims to analyze the influence of cognitive and affective factors on the adoption behavior of generative AI. Additionally, the study intends to explore the demographic characteristics of users and their impact on adoption behavior. The study's structure comprises a literature review, conceptual model development, cross-sectional survey design for data collection, analysis of results, and implications.

Literature Review and Hypothesis Development

Cognitive Factors: Perceived Ease of Use and Perceived Usefulness

TAM focuses on cognitive factors like perceived ease of use (PeoU) and perceived usefulness (PU) to explain the intention to adopt the new technology (Davis and Venkatesh, 1996). PEoU reflects users' beliefs about the effort required in using technology (Davis, 1989b; Marangunic´ and Granic´, 2015). Its importance in shaping attitudes and adoption intentions for new technologies (Dogra and Kaushal, 2023), particularly in the realm of generative AI platforms, is crucial (Maheshwari, 2023). High PEoU positively influences users' adoption intention, as it suggests ease in incorporating these platforms into their workflow (Prasad Agrawal, 2023; Wang and Zhang, 2023). PU, another critical factor, gauges users' belief in a technology's capability to enhance performance and productivity (Davis, 1989a). In the context of LLM-based generative AI, high PU correlates with increased adoption intention, as users recognize the tools' contributions to productivity and problem-solving (Ma and Huo, 2023). A harmonious interplay between PU and PEOU strengthens the positive usage intention toward LLM-based generative AI platforms, boosting the likelihood of adoption (Abdullah et al., 2016). Hence, relying on the above discussion, this study advanced the following hypotheses as follow:

H1: High PEoU of LLM-based generative AI platforms positively influence AI.
H2: High PU of LLM-based generative AI platforms positively influences the AI.
H3: PEoU of LLM-based generative AI platforms positively influences the PU.

Relative Advantage (RA)

CAT integrates affective factors like ARU and PL alongside the cognitive element RA

(Kulviwat et al., 2007). RA conceptualized as the extent to which technology is perceived as superior, highlighting efficiency gains and advancements in new technologies (Nasco et al., 2008). Identified as pivotal in adoption, users prefer technologies offering substantial improvements (Kulviwat et al., 2007; Rexfelt and Hiort af Ornäs, 2009). RA becomes crucial for efficiency gains over traditional technologies, influencing positive adoption intentions of new technology (Püschel et al., 2010). RA, positively interacting with Perceived Usefulness (PU) in CAT (Renko and Druzijanic, 2014), contributes to users' positive attitudes (Wang et al., 2011). Users acknowledging the streamlined and enhanced tasks facilitated by LLM-based generative AI perceive higher usefulness, emphasizing the interconnections of RA, PU, and the positive adoption intention of LLM-based generative AI platforms. Consistent with the above discussion, this study advanced the following hypotheses as follow:

H4: High RA of LLM-based generative AI platforms positively influence the PU.

H5: High RA of LLM-based generative AI platforms positively influence the AI.

Functional Value

TVC, Holbrook (2002) states that customers' or users' decisions on a product or service depend on their experience with the value generated by the product or service. Han et al. (2017) argued that FV is the most essential value when considering technology. In the realm of LLM-based generative AI platforms, the FV (Sweeney and Soutar, 2001) emerges as a crucial factor, representing the perceived utilitarian value derived from the experience of accomplishing task-related goals (Youn and Lee, 2019). As users engage with generative AI technologies, their perception of FV is intricately linked to two fundamental elements of the TAM: PEoU and PU (Rout et al., 2022). PEoU plays a significant role in shaping the user experience, as a user-friendly interface and intuitive interaction enhance the overall efficiency of task accomplishment, contributing positively to the FV (White et al., 2012). Concurrently, PU influences FV by determining how much technology, such as generative AI platforms, contributes meaningfully to goal attainment (Doshi et al., 2023). When users perceive these platforms as helpful in facilitating and improving task-related activities, the overall FV experienced is heightened. In line with previous studies, it can be noted that FV plays a pivotal role in shaping users' positive Intention toward adoption (Hong et al., 2017; Rout et al., 2022). Based on the above discussion, in the context of LLM-based generative AI, this study advanced the following hypotheses as follow:

H6: High PU of LLM-based generative AI platforms positively influence the FV.

H7: High PEoU of LLM-based generative AI platforms positively influence the FV.

Affective Factors: Pleasure and Arousal

In the realm of LLM-based generative AI platform adoption, affective factors such as pleasure and arousal play a crucial role in shaping users' intentions. Pleasure is the degree to which individuals experience positive feelings or satisfaction in a particular situation (Kulviwat et al., 2007), and becomes pertinent in generative AI platforms. Users who derive satisfaction and feel good about the technology, especially when experiencing its capabilities in enhancing efficiency and effectiveness, are more likely to exhibit a positive perception toward adoption (Kulviwat et al., 2007; Lee et al., 2011). Arousal, on the other hand, encompasses both physical activities and mental alertness triggered by surprising and novel stimuli (Mehrabian and Russell, 1974). Users may experience arousal when confronted with the technology's unprecedented speed, accuracy, and ability to perform relevant tasks efficiently (Cheng and Huang, 2022; Lee et al., 2012). Likewise, when LLM-based generative AI platforms perform relevant tasks efficiently, this may heighten attention and positive evaluation, resulting from generative AI's novel and effective nature and contributing to users' adoption intentions. Therefore, the affective factors of PL and AUR emerge as critical dimensions influencing the adoption dynamics of generative AI platforms. Hence, drawing on the CAT and the above discussion, this developed the hypotheses as follows:

H8: High expression of PL with LLM-based generative AI platforms positively influences the AI.

H9: High AUR generated with LLM-based generative AI platforms positively influences AI.

Mediating Role of Functional Value

The higher functional value (FV) corresponds to an increased intention to adopt new technology (Han et al., 2017). FV, in this context, encompasses the perceived utilitarian benefits users derive from interacting with LLM-based generative AI platforms (Hong et al., 2017). Building upon this premise, the study introduces FV as an intervening variable, mediating the relationship between PU and AI and PEoU and AI (Rout et al., 2022). The argument suggests that the impact of PU on adoption intention and the relationship between PEoU and adoption intention are mediated by users' perceptions of the FV offered by generative AI platforms. This mediation underscores the importance of users' subjective evaluations of the utilitarian benefits of LLM-based generative AI, providing an understanding of the intricate dynamics that influence adoption intention in the context of generative AI. Hence, this study proposed the following hypotheses as follow:

H10: High FV of LLM-based generative AI platforms positively influence the AI.

H11: The positive relationship between PU and the AI of LLM-based generative AI platforms mediated by FV.

H12: The positive relationship between PEoU and the AI of LLM-based generative AI platforms mediated by FV.

Methodology

This study adopts a cross-sectional and positivist research approach. A survey instrument was developed by borrowing per-validated scales from previous studies (Table 6.2). The items of scale used to measure PU, PEoU, FV, and AI in this research were anchored with an odd Likert Scale (7-Point), with one extreme representing strongly disagree (1-point) and another extreme representing strongly agree (7-point). The items used to measure two affective constructs, PL and ARU, were anchored with a 7-point emotion scale with extreme negative emotion (1-point) to extreme positive emotion (7-point). The self-administered questionnaire comprises two sections: one for demographic data collection and the other for capturing respondents' cognitive and affective states. In line with (Etikan et al., 2016), a purposive sampling technique was adopted to select the generative AI users. The respondents were selected based on their frequency of using LLM-based generative AI and the purpose of using generative AI. A person who used LLM-based generative AI at least four days a week and used it for professional tasks was the sample unit of this study. After setting the criteria, the qualifying participants were administered a questionnaire through an online platform. Data was collected from 2 January 2024 to 18 January 2024, and in between time frames, a total of 150 questionnaires were distributed and 107 correct responses were captured. Sample 107 forms a robust and adequate sample size for subsequent data analysis as the minimum sample size for the current study is 60, which is ten times

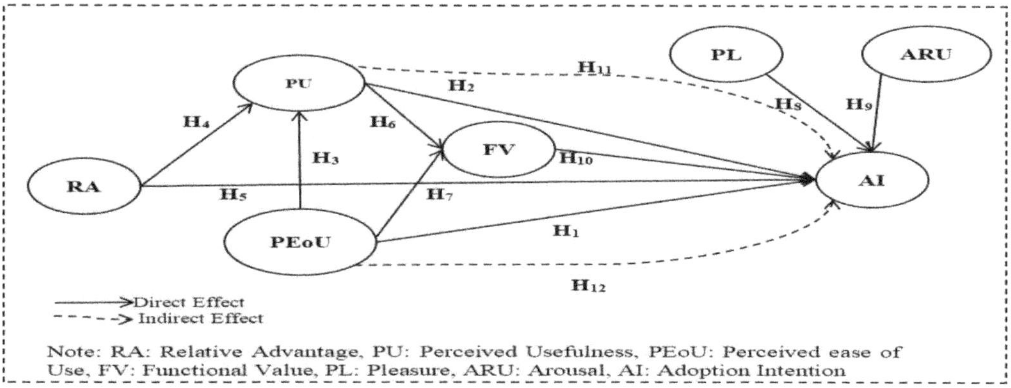

Figure 6.1 Proposed conceptual model
Source: Generated by Authors

the number of exogenous variables in the model (Hair et al., 2019; Rout et al., 2022). The study was conducted in Bhubaneswar, a city in India, and the sample encompasses both professionals and students, with a predominant representation of working professionals. This study assessed common method bias (CMB) using Harman's single-factor test, revealing a total variance of less than 50%, indicating no CMB (Podsakoff et al., 2003). For model testing (Figure 6.1), PLS-SEM is used. The Smart PLS 4.9 version facilitated PLS-SEM operations, while SPSS 27 was utilized for running the descriptive analysis.

Data Analysis

Generative AI users exhibit a diverse demographic profile (Table 6.1), showcasing inclusivity across genders and age groups. The technology attracts a balanced gender distribution and a significant portion of users aged between 25 to 31 years, indicating higher adoption rates among tech-savvy individuals. Additionally, active engagement from the 18 to 24 years age group suggests a broad appeal to the younger population. Educational qualifications vary among users, with a substantial percentage holding graduate and postgraduate degrees, emphasizing the technology's relevance across different academic backgrounds.

Regarding usage patterns, Chat GPT by Open AI is the most recognized and frequently used platform, with 100% awareness and usage among respondents. Bard by Google Inc. follows closely with high awareness and usage levels. Other platforms, like Claude by Anthropic, show lower awareness and usage percentages.

Users employ generative AI platforms for diverse purposes, with a focus on text generation (30%) and other tasks like blog writing, image processing, e-mail composition, and coding. This highlights the versatility of these platforms in catering to various needs. Overall, Chat GPT emerges as a prominent and widely utilized generative AI platform, reflecting its recognition, frequent usage, and diverse applications in tasks such as text generation, blog writing, and more.

Model Assessment

In the initial phase of PLS-SEM analysis, this study evaluates the outer model to ensure reliability and validity. The structural model (Inner Model) assessment uses the recommended 10,000 bootstrap samples (Streukens and Leroi-Werelds, 2016) to evaluate direct effects and mediating effects.

The measurement model underwent a rigorous evaluation to ensure reliability and validity. Factor loading of items, surpassing the recommended threshold of 0.708 (Hair

Table 6.1 Demographic profile and awareness and usage of generative AI platforms.

	Male	Female		
Gender (N: 107)	55(51.4%)	52(48.6%)		
	18-24	**25-31**	**32-38**	**39-55**
Age group (N:107)	25(23.7%)	55(51.4%)	18(16.8%)	9(8.4%)
	Graduate	**Postgraduate**	**PhD**	**Others**
Educational qualification (N:107)	48(42%)	38(36%)	10(9.3%)	11(10.2%)
	Brand name	**Awareness level(%)**		
Awareness about various generative AI platforms, brand name	Chat GPT (Open AI)	107(100%)		
	Bard (Google Inc)	70(56.4%)		
	Llama (Meta)	12(11%)		
	Gemini (Google Inc)	30(28%)		
	Claude (Anthropic's)	32(29%)		
	Microsoft Bing (Microsoft Corp.)	6(5%)		
		Frequently used(%)		
Frequently used generative AI platform	Chat GPT	107(100%)		
	Bard	104(97%)		
	Claude	15(14%)		
	Task	**Usage(%)**		
Purpose of use generative AI platform	Text Generation	32(30%)		
	Blog Writing	5(4%)		
	Image Processing	13(12%)		
	e-mail Writing	17(16%)		
	Coding	4(3.7%)		

Source: Generated by Authors

et al., 2019), confirmed indicator reliability (Table 6.2). Internal consistency and reliability were confirmed through Cronbach's alpha (α) values exceeding 0.7 (Hair et al., 2019) (Table 6.2) and CR values surpassing the same threshold of 0.7 (Hair et al., 2019) (Table 6.2). Additionally, all constructs demonstrated convergent validity, with AVE values exceeding 0.5 (Hair et al., 2019). Fronell-Larcker Criterion correlation matrix confirmed discriminant validity, where diagonal values surpassed the square

Table 6.2 Measurement model assessment.

Constructs	Items	Factor loading (λ)	VIF	Source
PU(α: 0.760, CR: 0.773, AVE:0.581)	PU1	0.752	1.406	
	PU2	0.849	2.058	
	PU3	0.722	1.807	Lund (2001)
	PU4	0.719	1.389	
	PU5	Dropped	Dropped	
PEoU (α: 0.906, CR: 0.937, AVE:0.727)	PEoU1	0.787	4.628	
	PEoU2	0.9	7.545	
	PEoU3	0.919	5.251	Lund (2001)
	PEoU4	0.942	7.134	
	PEoU5	0.69	4.649	
RA(α: 0.822, CR:0.823, AVE:0.740)	RA1	0.89	3.172	
	RA2	0.761	1.784	AlSaleh and Thakur (2019)
	RA3	0.879	2.200	
	RA4	Dropped	Dropped	
PL(α: 0.703, CR:0.720, AVE:0.641)	PL1	0.616	1.188	
	PL2	Dropped	Dropped	
	PL3	0.905	2.759	AlSaleh and Thakur (2019)
	PL4	0.851	2.468	
ARU(α:0.856, CR:0.950, AVE:0.871)	ARU1	0.907	2.277	AlSaleh and Thakur (2019)
	ARU2	0.958	2.277	
FV(α:0.856, CR:0.817, AVE:0.725)		0.878	3.801	
	FV1			
	FV2	0.917	4.053	Youn and Lee (2019).
	FV3	0.75	1.286	
AI(α:0.833, CR:0.834, AVE: 0.750)	AI1	0.91	2.651	
	AI2	0.853	1.915	AlSaleh and Thakur (2019)
	AI3	0.834	1.864	

Source: Generated by Authors

Note: The items were dropped due to low factor loading ($\lambda < 0.5$) and were mentioned as dropped. These low factors loading of items are dropped because, after deletion, the CR and AVE values of their respective constructs improved

roots of AVE in off-diagonals (Henseler et al., 2015) (Table 6.3). Preceding hypothesis testing, this phase addressed multi-collinearity concerns, revealing a maximum variance inflation factor (VIF) value of 7.545, below the recommended

Table 6.3 Fronell-Larcker criterion for discriminant validity.

	AI	ARU	FV	PEoU	PL	PU	RA
AI	***0.866***						
ARU	0.707	***0.933***					
FV	0.771	0.724	***0.851***				
PEoU	0.578	0.611	0.730	***0.853***			
PL	0.506	0.904	0.675	0.563	***0.801***		
PU	0.412	0.506	0.564	0.777	0.532	***0.762***	
RA	0.648	0.633	0.528	0.662	0.643	0.630	***0.860***

Source: Generated by Authors

Table 6.4 Structural model assessment.

Direct path model	**β**	**T-statistics**	**Decision**	**Mediating path model**	**β**	**T-statistics**	**Decision**
ARU->AI	1.174	5.278***	Accepted	FV->AI	0.683	10.134***	Accepted
PEoU->AI	0.138	1.355^{NS}	Rejected	PEoU->AI	-0.229	2.308**	Accepted
PEoU->FV	0.745	7.502***	Accepted				
PEoU->PU	0.641	7.490***	Accepted				
PL->AI	-0.854	4.357***	Accepted				
PU->AI	-0.113	1.212^{NS}	Rejected	PU->AI	-0.084	1.165^{NS}	Rejected
PU->FV	-0.014	0.125^{NS}	Rejected				
RA->AI	0.438	3.928***	Accepted				
RA->PU	0.205	2.554**	Accepted				
	R^2(AI): 0.870 R^2(FV): 0.533			PEoU->FV->AI	0.504	5.619***	Partial and competitive mediation
				PU->FV->AI	-0.007	0.083^{NS}	No mediation

Source: Generated by Authors

Note: NS: Not significant; **Significant (sig < 0.05); ***Significant(sig < 0.01)

threshold of 10 (Purwanto, 2021)and signifying the absence of potential multi-collinearity among the variables (Table 6.2).

Discussion

The objective of this study was to identify the cognitive and affective factors influencing the adoption intention of generative AI and empirically examine their impact on adoption intention. Drawing on TAM and CAT, this study identified PEoU, PU, and RA as cognitive factors and PL and AUR as affective factors. Building on the theory of value consumption, the study argued that FV is a crucial factor influencing adoption intention. The direct inner model assessment (Tabel 6.4) revealed that PEOU and

PU had no significant effect on the adoption intention of generative AI, challenging existing knowledge on TAM (Abdullah et al., 2016; Davis and Venkatesh, 1996; Rout et al., 2022), which asserts a positive impact of PU and PEOU on technology adoption behavior. The study suggested that users might not consider generative AI as a technology enhancing their skills but instead as a means of supporting them in performing tasks quickly and efficiently. The ease of use of generative AI may be perceived positively ($\beta = 0.641$, $p < 0.01$), influencing its usefulness, although managing it effectively might require additional effort and skill.

Furthermore, in line with (Kulviwat et al., 2014; Püschel et al., 2010), RA showed a substantial impact ($\beta = 0.438$, $p < 0.01$) on adoption intention, signifying that users perceive generative AI platforms as more beneficial than other technology support for specific tasks. Affective factor AUR positively influenced ($\beta = 1.174$, $p < 0.01$) adoption intention, indicating that users' excitement and stimulation lead to a higher adoption intention as generative AI makes tasks quicker, easier, and more accurate, and this finding is consistent with (Kulviwat et al., 2007). However, PL had a negative and significant ($\beta = -0.854$, $p > 0.01$) effect on adoption intention. The negative pleasure indicates that customers might be unhappy and annoyed with generative AI, which may diminish the adoption intention towards generative AI. and customers are unhappy and irritated due to the possible issues suggested by (Ivanov et al., 2024) like limited access, generating inaccurate information, reliance on outdated data, and ethical/security concerns contribute to user dissatisfaction and diminished adoption intention, which is contrary to the existing wisdom (Kulviwat et al., 2007; Nasco et al., 2008).

This study incorporated FV as a mediating factor between PEoU and, PU and AI. In the presence of FA, direct effect of PEoU on adoption intention is negative ($\beta = 0.438$, $p < 0.01$) but consistent with (Holbrook, 2002; Rout et al., 2022), the study found a strong mediating effect of FV($\beta = 0.504$, $p < 0.01$), suggesting that easiness and simple to use of generative AI platforms enhances users' perceived functional value, which refers that the easy to use and simple to learn to use generative AI help users to effectively complement their task and perform effectively, consequently increasing adoption intention. The mediating role of FV between PEoU and adoption intention indicated a strong impact of PEoU on FV. Despite the negative direct effect of PEoU in the presence of FV, the significant positive impact through FV suggested that a positive perception of generative AI's ease of use influences functional value, motivating users to adopt generative AI. Further, this study found a negative and non-significant impact of PU on adoption intention, and no significant mediating effect was found between PU, FV, and adoption intention. Previously, this study failed to establish any significant impact of PEoU and PU on adoption intention.

Conclusion

This study offers significant theoretical insights into human behavior toward generative artificial intelligence (AI). Integrating technology acceptance model (TAM), CAT, and the theory of value consumption constructs a model that reveals that traditional factors like perceived usefulness (PU) and perceived ease of use (PEoU) have no direct impact on generative AI adoption. However, a novel finding suggests a positive influence of PEoU when mediated by functional value (FV). Additionally, RA emerges as a crucial factor influencing the adoption of generative AI platforms, with both RA and PEoU impacting their perceived usefulness. Moreover, the study identifies two novel relationships between pleasure (PL) and arousal (AUR) with adoption intention, indicating their influence despite a negative impact from PL. Departing from conventional models, the study introduces FV as the immediate antecedent of adoption intention, proposing a novel competitive partial mediation among PEoU, FV, and adoption intention. Thus, it presents and validates a comprehensive model combining cognitive, affective, and value-based factors to elucidate generative AI adoption intention. The study suggests managerial actions for generative AI platform firms: enhance platform functionality with regular updates, focus on training models with updated data, prioritize speed and reasoning capabilities, ensure information authenticity, personalize data for emotional connections, and provide a competitive edge in the crowded market. For example, Google integrates large language model (LLM) AI in its search engine for personalized searches, while Microsoft incorporates Chat GPT in Excel for improved spreadsheet usage.

Limitations and Future Scope

This study's limitations include a small sample size, limited generalizability to other demographic groups, and a focus solely on working professionals and students in a single Indian region. Future research could expand the model to examine adoption behavior among entrepreneurs and MSMEs, increase the sample size for broader generalization, incorporate additional factors like social and economic value, ethical considerations, and perceived risk, introduce trust as a moderating variable, conduct longitudinal studies to assess pre and post-impacts and explore adoption behavior among different user groups through multi-group analysis. Future research can include demographic variables in their structural model to check the direct impact of these variables on the intention to adopt generative AI.

References

Abdullah, F., Ward, R., and Ahmed, E. (2016). Investigating the influence of the most commonly used external variables of TAM on students' perceived ease of use (PEOU) and perceived usefulness (PU) of e-portfolios. *Computers in Human Behavior*, 63, 75–90.

AlSaleh, D., and Thakur, R. (2019). Impact of cognition, affect, and social factors on technology adoption. *International Journal of Technology Marketing*, 13(2), 178–200.

Baek, T. H., and Kim, M. (2023). Is ChatGPT scary good? how user motivations affect creepiness and trust in generative artificial intelligence. *Telematics and Informatics*, 83, 102–130.

Cheng, L.-K., and Huang, H.-L. (2022). Virtual tourism atmospheres: the effects of pleasure, arousal, and dominance on the acceptance of virtual tourism. *Journal of Hospitality and Tourism Management*, 53, 143–152.

Chiu, T. K. F. (2023). The impact of generative AI (GenAI) on practices, policies and research direction in education: a case of ChatGPT and midjourney. *Interactive Learning Environments*, 1–17. doi:10.1080/10494820.2023.2253861.

Davis, F. D. (1989a). Technology acceptance model: TAM. *Information Seeking Behavior and Technology Adoption*, 205, 219.

Davis, F. D. (1989b). Perceived usefulness, perceived east of use, and user acceptance of information technology. *MIS Quarterly*, 13(3), 319–340. https://doi.org/10.1016/S0305-0483 (98)00028–0.

Davis, F. D., and Venkatesh, V. (1996). A critical assessment of potential measurement biases in the technology acceptance model: three experiments. *International Journal of Human-Computer Studies*, 45(1), 19–45.

Dogra, P., and Kaushal, A. (2023). The impact of digital marketing and promotional strategies on attitude and purchase intention towards financial products and service: a case of emerging economy. *Journal of Marketing Communications*, 29(4). 403–430.

Doshi, P., Nigam, P., and Rishi, B. (2023). Customer values and patronage intention in social media networks: mediating role of perceived usefulness. *VINE Journal of Information and Knowledge Management Systems*. https://doi.org/10.1108/VJIKMS-07-2022-0217

Etikan, I., Musa, S. A., and Alkassim, R. S. (2016). Comparison of convenience sampling and purposive sampling. *American Journal of Theoretical and Applied Statistics*, 5(1), 1–4.

Fui-Hoon Nah, Fiona, Ruilin Zheng, Jingyuan Cai, Keng Siau, and Langtao Chen. (2023). "Generative AI and ChatGPT: Applications, Challenges, and AI-Human Collaboration." *Journal of Information Technology Case and Application Research* 25 (3): 277–304. doi:10.1080/15228053.2023.2233814.

Gude, V. (2023). Factors influencing ChatGpt adoption for product research and information retrieval. *Journal of Computer Information Systems*, 1–10. https://doi.org/10.1080/08874417.2023.2280918

Hair, J. F., Risher, J. J., Sarstedt, M., and Ringle, C. M. (2019). When to use and how to report the results of PLS-SEM. *European Business Review*, 31(1), 2–24.

Han, L., Wang, S., Zhao, D., and Li, J. (2017). The intention to adopt electric vehicles: driven by functional and non-functional values. *Transportation Research Part A: Policy and Practice*, 103, 185–197.

Henseler, J., Ringle, C. M., and Sarstedt, M. (2015). A new criterion for assessing discriminant validity in variance-based structural equation modeling. *Journal of the Academy of Marketing Science*, 43, 115–135.

Holbrook, M. B. (2002). Introduction to consumer value. In Consumer Value. Routledge, pp. 17–44.

Hong, J.-C., Lin, P.-H., and Hsieh, P.-C. (2017). The effect of consumer innovativeness on perceived value and continuance intention to use smartwatch. *Computers in Human Behavior*, 67, 264–272.

Ivanov, S., Soliman, M., Tuomi, A., Alkathiri, N. A., and Al-Alawi, A. N. (2024). Drivers of generative AI adoption in higher education through the lens of the theory of planned behaviour. *Technology in Society*, 77, 102–521.

Kasneci, E., Seßler, K., Küchemann, S., Bannert, M., Dementieva, D., Fischer, F., et al. (2023). ChatGPT for good? on opportunities and challenges of large language models for education. *Learning and Individual Differences*, 103, 102–274.

Kulviwat, S., Bruner II, G. C., and Neelankavil, J. P. (2014). Self-efficacy as an antecedent of cognition and affect in technology acceptance. *Journal of Consumer Marketing*, 31(3), 190–199.

Kulviwat, S., Bruner II, G. C., Kumar, A., Nasco, S. A., and Clark, T. (2007). Toward a unified theory of consumer acceptance technology. *Psychology and Marketing*, 24(12), 1059–1084.

Lee, S., Ha, S., and Widdows, R. (2011). Consumer responses to high-technology products: product attributes, cognition, and emotions. *Journal of Business Research*, 64(11), 1195–1200.

Lee, W., Xiong, L., and Hu, C. (2012). The effect of facebook users' arousal and valence on intention to go to the festival: applying an extension of the technology acceptance model. *International Journal of Hospitality Management,* 31(3), 819–827.

Lund, A. M. (2001). Measuring usability with the use questionnaire12. *Usability Interface*, 8(2), 3–6.

Ma, X., and Huo, Y. (2023). Are users willing to embrace ChatGPT? exploring the factors on the acceptance of chatbots from the perspective of AIDUA framework. *Technology in Society*, 75, 102–362.

Maheshwari, G. (2023). Factors influencing students' intention to adopt and use ChatGPT in higher education: a study in the vietnamese context. *Education and Information Technologies*, 1–29. https://doi.org/10.1007/s10639-023-12333-z

Marangunić, N., and Granić, A. (2015). Technology acceptance model: a literature review from 1986 to 2013. *Universal Access in the Information Society*, 14, 81–95.

Mehrabian, A., and Russell, J. A. (1974). An Approach to Environmental Psychology. The MIT Press.

Nasco, S. A., Kulviwat, S., Kumar, A., and Bruner II, G. C. (2008). The CAT model: extensions and moderators of dominance in technology acceptance. *Psychology and Marketing,* 25(10), 987–1005.

Podsakoff, P. M., MacKenzie, S. B., Lee, J.-Y., and Podsakoff, N. P. (2003). Common method biases in behavioral research: a critical review of the literature and recommended remedies. *Journal of Applied Psychology*, 88(5), 879.

Prasad Agrawal, K. (2023). Towards adoption of generative AI in organizational settings. *Journal of Computer Information Systems*, 1–16. https://doi.org/10.1080/08874417.2023.2240744

Purwanto, A. (2021). Partial least squares structural squation modeling (PLS-SEM) analysis for social and management research: a literature review. *Journal of Industrial Engineering and Management Research,* 4 (2), 114–123.

Püschel, J., Afonso Mazzon, J., and Mauro C. Hernandez, J. (2010). Mobile banking: proposition of an integrated adoption intention framework. *International Journal of Bank Marketing*, 28(5), 389–409.

Renko, S., and Druzijanic, M. (2014). Perceived usefulness of innovative technology in retailing: consumers and retailersulness of innova. *Journal of Retailing and Consumer Services*, 21(5), 836–843.

Rexfelt, O., and Hiort af Ornäs, V. (2009). Consumer acceptance of product-service systems: designing for relative advantages and uncertainty reductions. *Journal of Manufacturing Technology Management*, 20(5), 674–699.

Rout, K., Sahoo, P. R., Bhuyan, A., Tripathy, A., and Smrutirekha (2022). Online grocery shopping behavior during COVID-19 pandemic: an interdisciplinary explanation. *Cogent Business and Management,* 9(1), 2084969.

Sætra, H. S. (2023). Generative AI: here to stay, but for good?. *Technology in Society,* 75, 102–372.

Streukens, S., and Leroi-Werelds, S. (2016). Bootstrapping and PLS-SEM: a step-by-step guide to get more out of your bootstrap results. *European Management Journal*, 34(6), 618–632.

Sweeney, J. C., and Soutar, G. N. (2001). Consumer perceived value: the development of a multiple item scale. *Journal of Retailing*, 77(2), 203–220.

Tiwari, C. K., Bhat, M. A., Khan, S. T., Subramaniam, R., and Khan, M. A. I. (2023). What drives students toward ChatGPT? an investigation of the factors influencing adoption and usage of ChatGPT. *Interactive Technology and Smart Education.* https://doi.org/10.1108/ITSE-04-2023-0061

Wang, Y., and Zhang, W. (2023). Factors influencing the adoption of generative AI for art designing among chinese generation Z: a structural equation modeling approach. *IEEE Access*, 11, 143272–143284

Wang, Y., Meister, D., and Wang, Y. (2011). Reexamining relative advantage and perceived usefulness: an empirical study. *International Journal of Information and Communication Technology Education (IJICTE)*, 7(1), 46–59.

White, C., McMurray, A. J., and Rudito, P. (2012). Using mature concepts to generate new ideas: technology acceptance revisited. *International Journal of Technology Marketing*, 7(4), 361–378.

Youn, Song-yi, and Kyu-Hye Lee. (2019). "Proposing Value-Based Technology Acceptance Model: Testing on Paid Mobile Media Service." Fashion and Textiles 6 (1): 13. https://doi.org/10.1186/s40691-018-0163-z.

7 Sustainable leadership and its effects on organizational change: a theory development, context, characteristics, and methodology framework based bibliometric analysis

Samir Kumar Mahapatro[a], *Gayatri Panda*[b] *and Akankshya Patnaik*[c]

Department of Management, NIST University, Berhampur, Odisha, India

Abstract

In today's dynamic socioeconomic setting, effective leadership is crucial to maintaining the sensitive equilibrium between prosperity, social justice, and protecting the environment. Sustainable leadership, prioritizing people, the environment, and profit, has emerged as a fundamental paradigm. It highlights leaders' critical role in directing their teams through the changes required to accomplish triple bottom-line targets, i.e., social, environmental, and economic targets. The impact of sustainable leadership on organizational change is examined in this hybrid review using an innovative approach that combines bibliometric analysis and the theory development, context, characteristics, and methodology (TCCM) framework analysis. A detailed search of the Scopus database has gathered multiple articles to answer relevant research questions: 1. tracing the evolution of sustainable leadership and change management, and 2. concerning dominant theories, contexts, characteristics, and methodologies to identify gaps and suggest future pathways. The review shows that effective change management strategies and sustainable leadership practices are positively correlated, which improves organizational performance. This review also paves the way for future researchers and scholars to study the topic and contribute more to the area of research. This review provides new insights and perspectives on the subject through the innovative integration of bibliometric analysis and the TCCM framework analysis.

Keywords: Change management, ethical leadership, organizational change, planned change, sustainable leadership

Introduction

Organizations must combine financial goals with social responsibility and environmental impact considerations for sustainable growth in the dynamic global landscape that is being changed by globalization and ecological dynamics (Iqbal et al., 2020). Long-term success in sustainable development depends on stakeholders adopting new perspectives and navigating difficulties (Nandedkar et al., 2023; Piwowar-Sulej and Iqbal, 2023). Companies that want to contribute significantly to promoting sustainable practices must understand how transformation and sustainability are intertwined (Tipu, 2022). Organizational change is essential because of the need for innovative thinking and the interaction of internal and external forces (Jeronen, 2023). Law, politics, society, religion, ethics, and culture must all be considered when transitioning to business sustainability. Businesses that want to create long-lasting change must enable people and groups to overcome discomfort. When faced with politicized

[a]Mahapatro.samir1991@gmail.com, [b]gayatri@nist.edu, [c]akankshya@nist.edu

DOI: 10.1201/9781003564447-7

transformation projects, leadership in large enterprises demands bravery (Stead and Stead, 2014). Kurt Lewin's three phases of transformation are the basis for effective change management, which depends on visionary leadership that emphasizes vision, involvement, and stakeholder consideration (Bolden, 2016).

In the modern business landscape, leadership plays an undeniable role as guides, unifiers, inspirers, and change catalysts, driving collective goals forward. While traditional leadership models have proven effective in navigating organizational and societal complexities, the current environment necessitates a shift (Hunger, 2020; Graetz, 2000). Today, amidst pollution's reach, economic volatility, and declining quality of life, a new leadership paradigm must emerge—one capable of addressing multifaceted challenges comprehensively, and therefore, sustainability leadership emerged. These leaders foster environments where individuals embark on enlightening journeys, crafting solutions to sustainability dilemmas. Unlike traditional top-down approaches, they prioritize collaboration and adaptability in action planning, acknowledging the ever-changing landscape (Asbari et al., 2021).

In contrast to traditional leadership's focus on profitability at the expense of employee well-being and societal and environmental considerations, sustainability leadership embodies a holistic approach. Autocratic, democratic, transactional, and transformational leadership styles, once dominant, find themselves less effective in today's dynamic context. According to (Leal Filho et al., 2020), visionary leadership fosters collaboration, coordination, and sustainable development. Sustainable leadership is vital to contemporary organizational studies because it prioritizes quality of life and societal collaboration (Leal Filho et al., 2019). Environmental concerns, social responsibility, changing economic landscapes, and modern corporate paradigms drive scholarly interest in leadership's role in sustainability and change (Silvestri and Veltri, 2020). Regardless of their organizational rank, change agents with the guts to support business sustainability efforts frequently take on the role of sustainability leaders (Acton, 2021), highlighting the undeniable importance of leadership in inspiring teams to embrace change for long-term, sustainable development.

Very few researchers have tried to review the available articles on organizational change management and sustainable leadership through rigorous literature studies. For instance, (Rehman et al., 2019) carried out a systematic review of sustainable leadership in the context of public sector organizations, and (Attah, 2017) explored the fundamentals of sustainable leadership. However, the research on sustainable leadership practices and change management seems fragmented and incomplete despite these attempts to compile the specialized literature in these areas. This disparity emphasizes how crucial it is to comprehensively study sustainable leadership's past, present, and future to manage organizational change effectively. The main aim of the current study is to fill this gap by combining a bibliometric and theory development, context, characteristics, and methodology (TCCM) framework methodology with a literature analysis to provide a thorough overview of the topic. This study is unique because both methods review sustainable leadership and change management literature. Top of form The research addresses three key research questions:

1. What journals, countries, and publication years have made the most substantial contributions to this subject, along with the most cited articles?
2. In what ways do various network analyses, coward analyses, keyword co-occurrence analyses, and cluster analyses elucidate distinct realms of understanding?

3. What theories, contexts, characteristics, and methods are applied, identifying gaps and suggesting future directions?

Brief Overview of Sustainable Leadership and Change Management

The World Commission on Environment and Development's Brundtland Report, which highlighted the interwoven pursuit of social, environmental, and economic welfare (Bansal and DesJardine, 2014; Hajian and Kashani, 2021; Kutlu, 2020), gave rise to the idea of corporate sustainability. The old profit-driven "business-as-business" strategy is challenged by corporate sustainability, which involves integrating these three components into strategic organizational management. These strategic forces fundamentally change objectives and vision (Muff et al., 2020). The meaning of "sustainability leadership" has changed in organizational and change management literature over the past 15 years, and this has been acknowledged in talks about the implications of business sustainability for organizational reform across various industries and analytical levels (Leal Filho et al., 2020). Although some claim that senior management roles are the primary role of sustainability leaders, the strategy-as-practice approach acknowledges leaders not part of the top management group, such as middle managers and technical specialists (Hargreaves and Fink, 2012; Oriade et al., 2021). By taking measured risks, sharing knowledge, and comprehending their place in interrelated systems, sustainability leaders can significantly contribute to a sustainable future. According to (Metcalf and Benn, 2013), effective sustainability leadership necessitates managing complexity, considering various viewpoints, and integrating everyone in dynamic organizational evolution.

In the contemporary global corporate context, where paradigms constantly change, leaders must modify their perspectives and demonstrate humility, stressing the significance of adopting sustainability principles (Ferdig, 2020). Beyond typical managerial achievements, leadership for sustainability necessitates a readiness to pick up new abilities and function in a turbulent marketplace. To address sustainability challenges, sustainable leaders must be humble and dedicated to fostering inclusivity, cooperation, and win-win partnerships—a quality that transcends self-interest (Myers and Rushbrook, 2024; Akcay Kasapoglu, 2018; Nguyen et al., 2021). Recognizing their ability to bring about positive change and alter the world, these leaders embrace uncertainty and change (Ferdig, 2020). To adopt creative strategies that promote Innovation, lower employee turnover, and produce long-term profits, sustainable leadership acknowledges the necessity of change and takes measured risks (Guzmán et al., 2020; Oberer and Erkollar, 2018). The study emphasizes how change-embracing sustainable leadership affects transformation and, as a result, organizational performance, and the study evolved.

Research Methodology

The author employed a systematic bibliometric review methodology and the TCCM framework in this review. The Scopus database was extensively searched for relevant literature due to its broader coverage of literature on management studies compared to the Web of Science (Singh et al., 2021; Pranckutė, 2021). The Scopus search was conducted using the given parameters: (Title-ABS-key ("Sustainable leadership" or "Ethical leadership" or "Responsible leadership" or "Eco-friendly" or "Green leadership" or "Collaborative leadership") and Title-ABS-key ("Organizational Change" or "Transformational change" or "Structural reform" or "Structural Change" or "Adaptive change"). The result initially had 111 articles. The author applied strict filters, considering only articles and reviews published in journals or book chapters. These filters resulted in a final selection of 24 English-language papers on sustainable leadership and change management published between 2014 and 2023 which can be clearly seen in Figure 7.1. The TCCM framework emphasizes theory development, context, characteristics, and methodology and summarizes and builds theories from the research articles (Shaik and Dhir, 2020). Extracted data from the articles, including author(s), title, year of publication, source title, applied theories, the context of the study, characteristics focus of the study, and detailed method used in the study, were systematically organized in a table to showcase the Framework clearly (Zhu and Liu, 2020). For statistical analysis and clustering, the author used RStudio, a user-friendly tool based on the R programming language and VOS viewer (Martín-Martín et al., 2021).

Analysis

Countries' scientific production and Trend of publication over time: Countries with high publication contributions are highlighted in Figure 7.2, which includes the USA, Portugal, Canada, Australia, the UK, China, Lithuania, and Pakistan. India's contribution is lower, indicating that more attention is required. The annual scientific productions are shown in Figure 7.3, focusing on the peaks in 2017, 2021, and 2023. Despite ups and downs, the topic draws scholarly attention, indicating its ongoing appeal and significance in this 21st-century world.

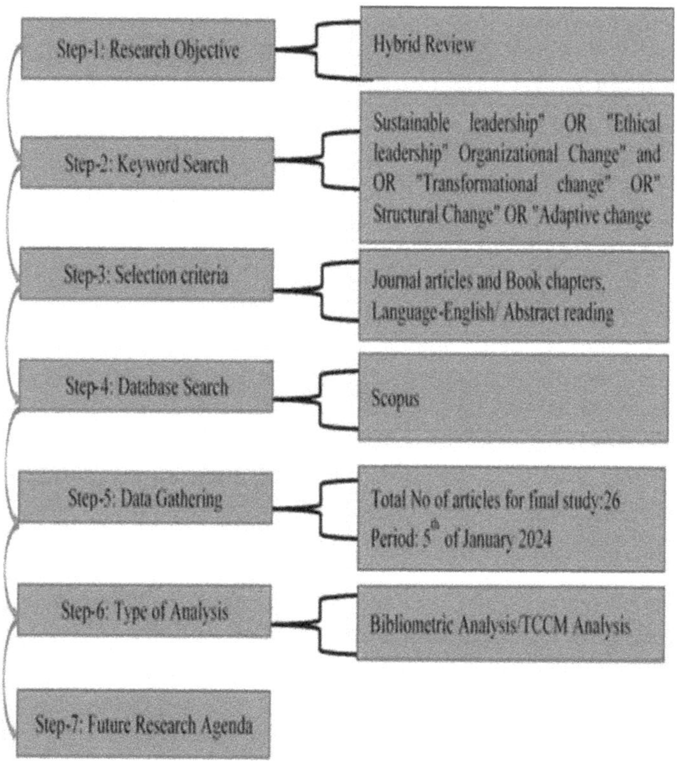

Figure 7.1 Seven-step methods
Source: Authors Compailation

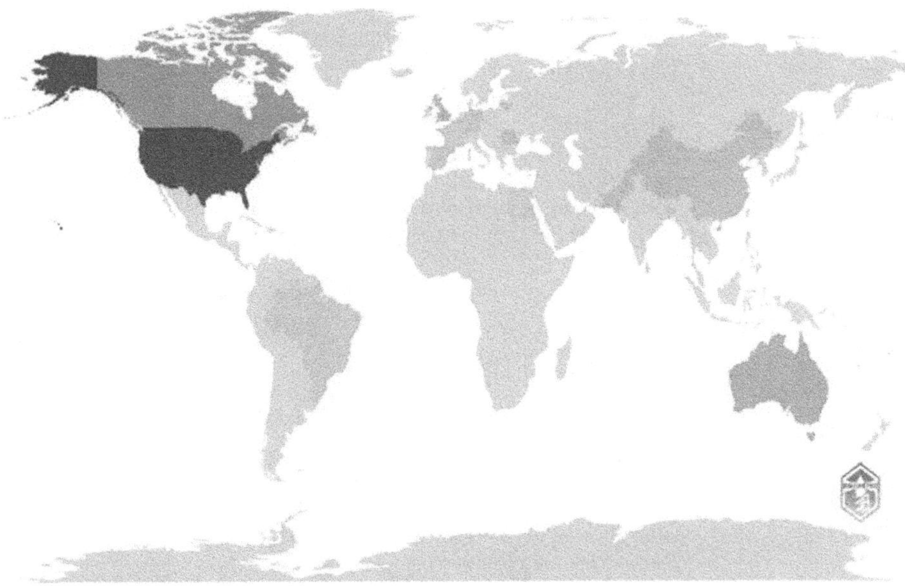

Figure 7.2 Countries scientific production
Source: Authors Compailation

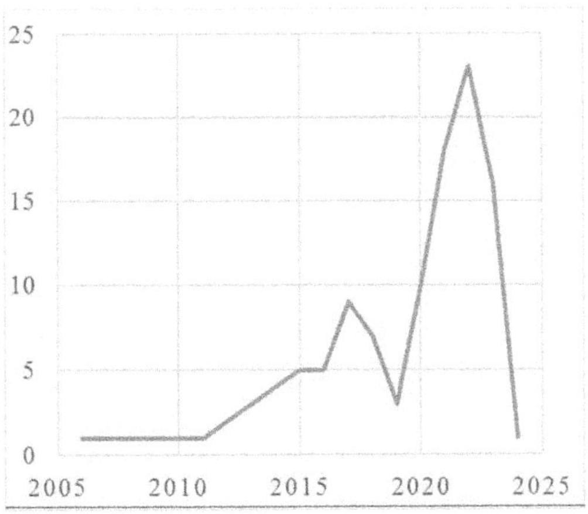

Figure 7.3 The Trend of publication over time
Source: Authors Compailation

As with every scientific investigation, some foundational works serve as essential stimulants, driving the field's advancement and determining its course. The importance of these seminal works in the field of sustainable leadership and change management cannot be emphasized. Table 7.1 provides a thorough overview of the major publications that have had a significant impact on the study by listing the top five papers, top five journals in order of citation count.

Keyword Co-occurrence Analysis
The keyword co-occurrence analysis makes it easy to determine which subjects are gaining momentum in a given field and identify potential future research fields.

Keywords reveal much about the subject matter of articles, and the author used VOS viewer to do a keyword co-occurrence analysis (Saha et al., 2020). The study of the frequency with which specific keywords appear in the articles we looked at is shown in Figure 7.4 below. The ones that show up a lot are 'Organizational Change' (15-OC and 35-TLS), 'Leadership' (12-OC and 45-TLS), 'Ethical leadership' (11-OC and 19-TLS), 'Collaborative Leadership' (07-OC and 15-TLS), 'Ethics' (04-OC and 14-TLS), 'Transformational Leadership' (04-OC and 12-TLS), 'Sustainability' (04-OC and 12-TLS), 'Sustainable development'

Table 7.1 Top five journals and highly cited papers.

Top five journals		Top five highly cited articles		
Sources	**Articles**	**Article name**	**Authors**	**TC**
journal of applied behavioral science	2	Institutional change and leadership associated with blended learning innovation: Two case studies	(Garrison and Vaughan 2013)	391
Journal of business ethics	2	Do erceptions of Ethical Conduct Matter During Organizational Change? Ethical Leadership and Employee Involvement	(Sharif and Scandura 2014)	321
Sustainability (Switzerland)	2	Eco-innovation, responsible leadership, and organizational change for corporate sustainability	(Paraschiv et al. 2012)	158
Amfiteatru economic	1	How Ethical Leadership Shapes Employees' Readiness to Change: The Mediating Role of an Organizational Culture of Effectiveness	(Metwally et al. 2019)	155
Others	19	The ethics of conscious capitalism: Wicked problems in leading change and changing leaders	(Fyke and Buzzanell 2013)	141

Source: Authors Compailation

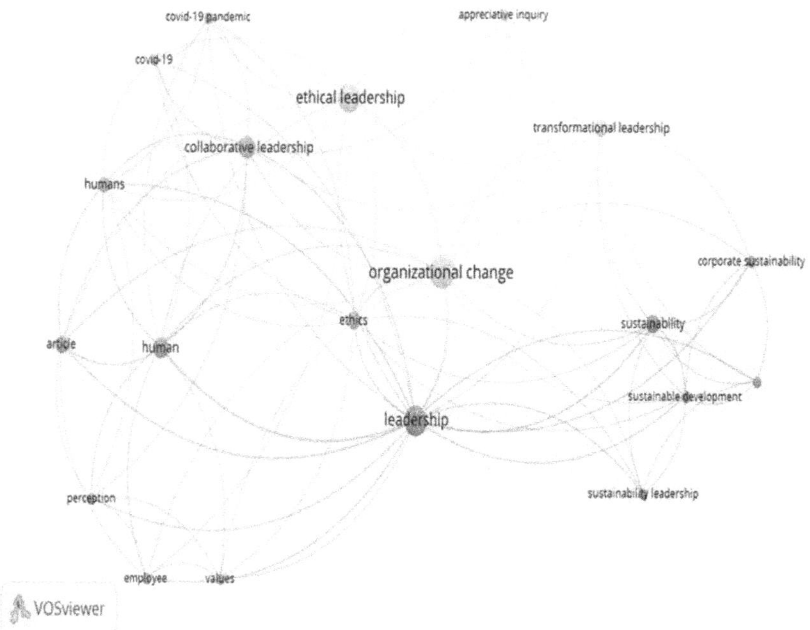

Figure 7.4 Keyword co-occurrence analysis
Source: Authors Compailation

(02-OC and 10-TLS), and 'Sustainable Leadership' (02-OC and 07-TLS). These words help us understand the current research topics, allowing us to plan future study patterns. Other researchers who wish to advance this topic may also find this material valuable. This analysis shows that extensive research has been conducted on various leadership styles and organizational change. Consequently, the authors have identified a particular interest in researching the topic of sustainable leadership and its effects on managing organizational change.

(*OC-Keyword co-occurrences, *TLS-Total link strength)

Co-word Analysis

This review of Four research publications on leadership and change management found 14 authors' keywords and eight indexed keywords using a minimum occurrence of two keywords as a criterion. Investigating these phrases in both sources reveals that the authors emphasized various types of leadership in handling different aspects of change for the organization's development. Conversely, indexers emphasized the leadership's role in improving human well-being as mentioned in the Figure 7.5.

Cluster Analysis

Cluster 1: Leadership for sustainability development: This cluster explores leadership approaches and strategies that promote sustainability and sustainable development. It may involve examining how leaders can drive organizational initiatives prioritizing environmental, social, and economic sustainability.

Cluster 2- Aligning ethics and values for organizational change: This cluster explores the importance of aligning ethical principles and values with organizational change efforts. It may involve studying how ethical considerations and value systems influence decision-making processes during organizational change and how leaders can ensure that changes are implemented consistent with the organization's ethical Framework.

Cluster 3-Ethical Leadership and Resilience in the COVID-19 Crisis: This cluster examines how ethical leadership practices contributed to organizational resilience during the challenges posed by the COVID-19 pandemic. It explores how leaders can navigate ethical dilemmas while effectively guiding their organizations through the crisis. Figure 7.6.

TCCM analysis: bridging the gap and guiding future research

The TCCM framework becomes essential to understand sustainable leadership and change management better and pave the way for future research. The framework, known as TCCM," was used to pinpoint the gaps in the existing literature. In simpler terms, TCCM is explained in a table (Table 7.2).

Theory Development: Evolving Perspectives

It is clear from a thorough analysis of 24 papers that there are many distinct viewpoints within the field of leadership theories, including behavioral theories, change management frameworks, and various leadership styles. Still, there needs to be more thorough literature on these topics. In a time when things are constantly changing, sustainable leadership is still in its infancy and has much potential for research and development. This study uses the TCCM framework as a guiding tool to conduct a literature review to close this knowledge gap. The main objective is to stimulate the development of new ideas that can clarify the complexities of sustainable leadership in the context of change management. Although current theories offer a fundamental structure, they require empirical examination and improvement in subsequent studies.

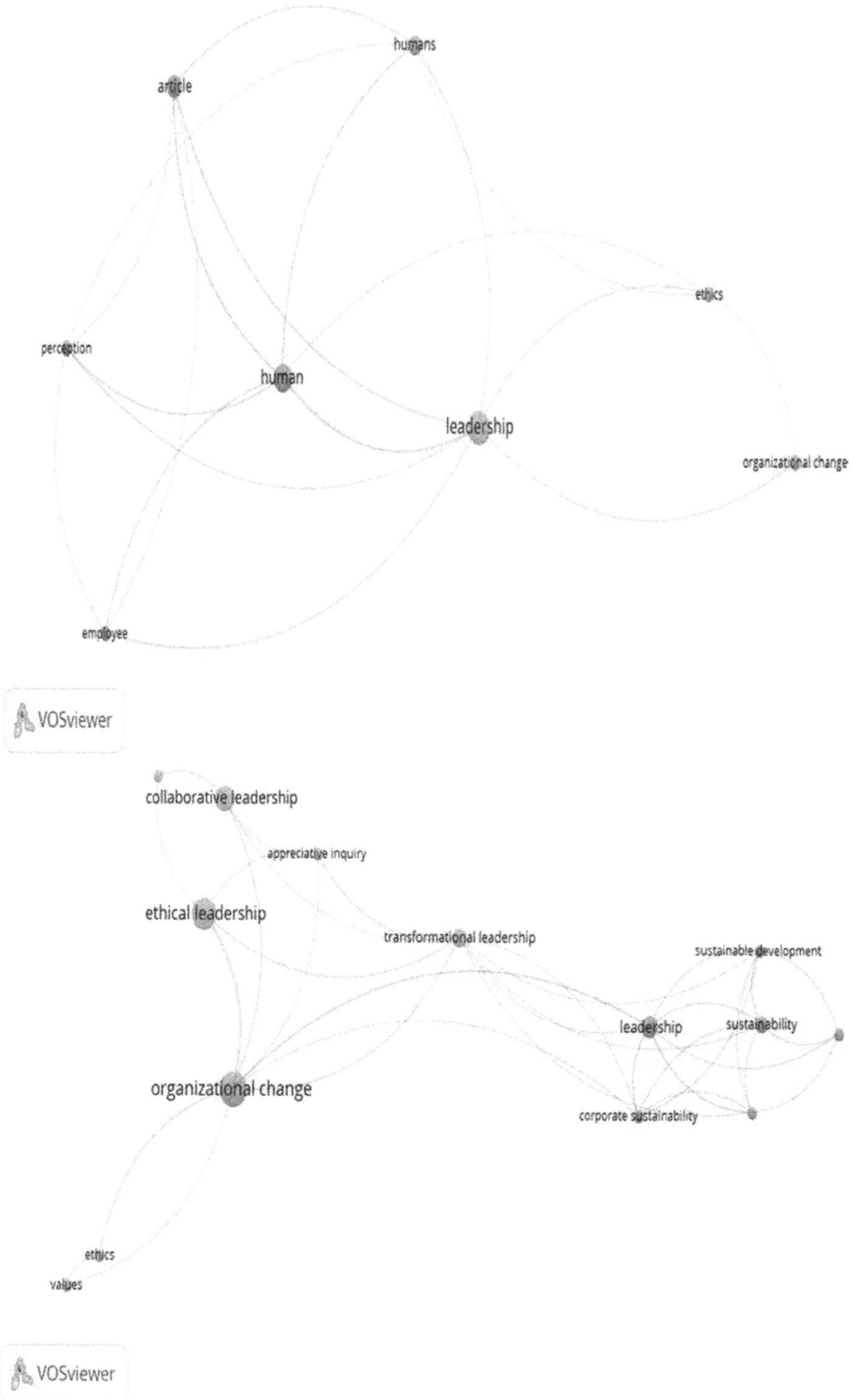

Figure 7.5 Co-word analysis (index keyword vs. author keyword cooccurance analysis)
Source: Authors Compailation

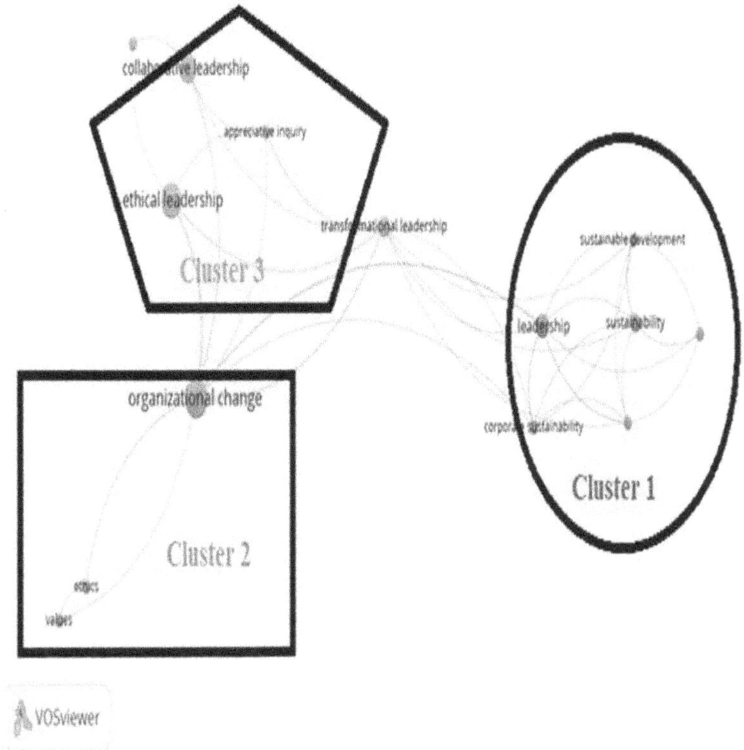

Figure 7.6 Cluster map
Source: Authors Compailation

Table 7.2 TCCM analysis.

1	**Article name (year)**- Leaders as coaches: toward code of ethics-2023 **Theory/concept**- Ethical leadership theory, Stakeholders theory **Context**- Coaching leadership of organizational betterment. **Characteristics**- The study pinpoints seven ethical concerns in leader-coach relationships: definition ambiguity, conflict of interest, confidentiality, power imbalance, freedom to participate, boundaries, and favoritism. **Methodology**- Critical incident method.
2	**Article name (year)**- The relationship between principals' perception of their leadership style and positive attitudes toward change in the light of the COVID-19 pandemic -2023. **Theory/concept**- Leadership styles theories, change management theory, and Cognitive dissonance theory. **Context**- Leadership in crisis management **Characteristics**- This study analyzes specific leadership that contributes to a positive orientation and adaptability in the face of pandemic-related challenges. **Methodology**- Survey method (qualitative method) statistical analysis: correlation method
3	**Article name (year)**- I value ethics, do you? An employee-oriented perspective on ethical leadership and organizational change -2023 **Theory/concept**- Ethical leadership theories and change management frameworks. **Context**- Ethical leadership for organizational change. **Characteristics**—The keys to success are Aligning ethical leadership, boosting employee engagement, and integrating ethics into strategic decisions. **Methodology**- Surveys and quantitative assessments.

4	**Article name (year)**- Prevailing leadership styles in change management: evidence from existing research. -2023. **Theory/concept**- Leadership styles theory and Change management theory. **Context**- Transformational leadership for organizational transformation. **Characteristics**-Change leadership styles include transformational inspiration, transactional rewards/punishments, charismatic influence, and situational adaptability. **Methodology**- Literature reviews, meta-analysis, and case study analysis.
5	**Article name (year)**- Collaborative leadership to support sustainability in practice for dietitians as allied health professionals -2023 **Theory/concept**- Transformational leadership, distributed leadership, or systems theory **Context**- Collaborative leadership in the healthcare system. **Characteristics**- Effective communication, teamwork, and integrating sustainability principles into daily practice can help achieve corporate sustainability. **Methodology**- Survey and semi-structured interview (qualitative and quantitative method)
6	**Article name (year)**- An informed practice for leading change collaboratively -2022 **Theory/concept**- Change management theory, leadership theories, Behavioral theories **Context**- Collaborative leadership for development. **Characteristics**- Strategic thinking, effective communication, adaptability, and a commitment to ongoing learning can be helpful for an informed change. **Methodology**- Qualitative analysis
7	**Article name (year)**- Creating a sustainable model for stroke system change. **Theory/concept**- Innovations theory, systems theory, and sustainability theory. **Context**- Leadership for sustainable system development. Characteristics such as stakeholder engagement, scalability, adaptability to evolving medical technologies, and long-term funding mechanisms are needed to craft a sustainable change model. **Methodology**- Wenger's phases of value creation and qualitative and quantitative theory
8	**Article name (year)**- Linking ethical leadership to employees' change-oriented Organizational Citizenship behavior: a multilevel moderated mediation model -2022 **Theory/concept**- Social exchange theory and cue consistency theory **Context**- Ethical leadership for organizational change management **Characteristics**- organizational citizenship behavior, ethical leadership, and organizational justice for overall betterment. **Methodology**- In-depth interviews (qualitative analysis) statistical tools: hierarchical linear model
9	**Article name (year)**- The effect of trait anger and impulsiveness on ethical leadership and support for organizational change -2021 **Theory/concept**- Ethical leadership theory and trait theory. **Context**- Effect traits of leadership on organizational change. **Characteristics**- Ethical leadership, personality, and leadership traits are needed for organizational growth. **Methodology**- Survey (qualitative method) statical tools: regression analysis
10	**Article name (year)**- A crucial moment for reflection on the importance of ethical leadership in academic medicine -2021 **Theory/concept**- Ethical leadership theory **Context**- Ethical leader for discrimination management. **Characteristics**—Leaders navigate extreme disruptions, test principles, loyalties, and responsibilities, and demand ethical decision-making for short—and long-term solutions. **Methodology**- Review study and case study analysis.
11	**Article name (year)**- Ethical leadership in times of change: the role of change commitment and change information for employees' dysfunctional resistance -2021 **Theory/concept**- Ethical leadership, social exchange, and change management theories. **Context**- Ethical leadership influences change commitment.

Characteristics- Ethical leaders contribute to positive change commitment, diminishing dysfunctional resistance. Employees' perception of change information moderates the relationships, influencing the strength of the effects.
Methodology- Survey (qualitative and Quantitative method)

12 **Article name (year)**- An evidence-based approach to academic management in a School of Public Health using the Smart Model-2021
Theory/Concept- Organizational behavior theory and educational leadership theory.
Context- Leadership for better academic management.
Characteristics- The SMAART SPH dashboard fosters transparency, equity, and informed decision-making.
Methodology- Survey and data analysis.

13 **Article name (year)**- Global leadership for sustainability -2021
Theory/concept- Spiritual leadership theory and Sustainability theory.
Context- Spiritual leadership for global sustainability.
Characteristics- GLFS integrates spirituality, emphasizing self-transcendence and interconnectedness. Leaders prioritize a triple bottom line and balance stakeholders' demands.
Methodology- Literature reviews.

14 **Article name (year)**- State-of-the-art practices being reported by the PRME champions group: a reference to advance education for sustainable development-2020
Theory/concept- Sustainable development theory, Stakeholders theory, Educational leadership theory.
Context- Educational leadership for advanced education management.
Characteristics- The resulting Framework guides educational institutions in effecting transformational change aligned with responsible management principles.
Methodology- Institutional document analysis (Qualitative method)

15 **Article name (year)**- Development and application of a distributed leadership framework to the curriculum mapping of quantitative skills in first-year undergraduate agriculture degrees -2019
Theory/concept- Distributed leadership theory and Organisational learning theory.
Context- Distributive leadership for skill and knowledge development.
Characteristics- A modified distributed leadership model incorporates change theory principles to drive curriculum mapping. The emphasis is on improving quantitative skills development in agriculture degrees, with curriculum maps pinpointing when these skills are taught, practiced, and assessed across various science programs.
Methodology- Literature review and case study analysis.

16 **Article name (year)**- How ethical leadership shapes employees' readiness to change: the mediating role of an organizational culture of effectiveness -2019
Theory/concept- Ethical leadership theory, Change management theory, and Organizational culture theory.
Context- Leadership for effective organizational management.
Characteristics- Ethical leaders positively influence employee readiness to change by reducing uncertainty. The study investigates the mediating role of an organizational effectiveness culture in this relationship.
Methodology- Survey (qualitative analysis)

17 **Article name (year)**- How the circle model can purpose-orient entrepreneurial universities and business schools to serve society? -2017 truly
Theory/concept- Social development theory, Innovation, and development theory.
Context- Entrepreneurial universities to transcend organizational and social effectiveness
Characteristics- The circle model extends the role of entrepreneurial universities to encompass education, research, and organizational purpose, fostering a developmental perspective. The focus is on transforming structures and strategies to serve societal well-being.
Methodology- Literature reviews.

18	**Article name (year)**- Beyond unsustainable leadership: critical social theory for sustainable leadership -2017 **Theory/concept**- Social development, Sustainable leadership, and system thinking theories. **Context**- Leadership for sustainable development. **Characteristics**- Critical leadership studies provide valuable insights for researching leadership in sustainable development. The paper critiques mainstream leadership, emphasizing the limitations of existing paradigms. **Methodology**- Literature reviews.
19	**Article name (year)**- A way forward: cascading ethical and change leadership, values enactment, and group-level effects on commitment in corruption recovery -2017 **Theory/concept**- Ethical leadership, values enactment, and crisis management theories. **Context**- Leadership for ethical organizational practices. **Characteristics**-Ethical leadership and values enactment positively impact organizational commitment post-corruption. Shared employee perspectives on business unit leaders' ethical leadership influence commitment to help reduce corruption. **Methodology**- Survey method (qualitative and quantitative).
20	**Article name (year)**- From start-up to sustainability: a decade of collaboration to shape the future of nursing -2017 **Theory/concept**- Organizational learning theory, sustainability theory. **Context**- Leadership for health care system sustainability. **Characteristics**- The consortium's evolution focuses on improving efficiency and sustainability in organizational infrastructure. It discusses shared curriculum delivery to enhance accessibility and affordability in nursing education. **Methodology**- Literature reviews.
21	**Article name (year)**- Ethical leadership's potential and boundaries in organizational change: a moderated mediation model of employee silence-2016 **Theory/concept**- Ethical leadership, Mediation, and employee development theories. **Context**- Ethical leadership for affective commitment to change. **Characteristics**- The study examines four forms of employee silence (acquiescent, quiescent, prosocial, and opportunistic) as mediators between ethical leadership and commitment to change. It also explores the moderating roles of political perceptions and personal change impact. **Methodology**- Survey (quantitative method)
22	**Article name (year)**- Locating sustainable leadership within a typology of leadership in business -2016 **Theory/Concept**- Stakeholders theory and Sustainable leadership theory **Context**- Leadership for corporate sustainability. **Characteristics**- The paper identifies four types of leaders and aligns them with the business context where their approach may be suitable. It aims to place sustainable leadership within a typology that correlates leadership styles with environmental beliefs. **Methodology**- Literature reviews.
23	**Article name (year)**- Do perceptions of ethical conduct matter during organizational change? Ethical leadership and employee involvement-2014 **Theory/concept**- Ethical leadership and Change management theory and employee development theory. **Context**- Ethical leadership in organizational change and its combined impact on work-related attitudes and performance. **Characteristics**- Ethical leadership's impact is examined alongside organizational change and involvement in change as moderators. **Methodology**- Survey method (quantitative approach).
24	**Article Name(Year)**- Employee satisfaction and sustainable leadership practices in Thai SMEs-2014 **Theory/concept**- Sustainable leadership theory, organisation development theory, and social exchange theory.

Context- Effect of sustainability leadership practices on employee satisfaction.
Characteristics- The study identifies 20 out of 23 sustainable leadership practices significantly linked to enhanced employee satisfaction. Key predictors include high staff engagement, valuing employees, ethical behavior, organizational change considerations, a strong vision, an enabling culture, and product/service quality.
Methodology- Survey (qualitative method)

Source: Authors Compailation

Context: Expanding Horizons

Previous studies mainly focused on the role of various leadership styles in performance management, change management, crisis management, and organizational development. TCCM analysis emphasizes the need for more studies in multiple contexts to examine how sustainable leadership manages organizational change in various sectors and countries. The study emphasizes the need to understand how sustainable leadership impacts organizational change, a concept primarily ignored in previous studies.

Characteristics: Unpacking Constructs

This study delves into constructs like sustainable leadership and change management. It examines various leadership styles, including transformational, ethical, and sustainable leadership, and their impact on organizational change, employee performance, and overall well-being. The analysis considers antecedents such as adaptability, ethical decision-making, long-term vision, and innovation. Future studies should build on these concepts, adding dimensions and exploring new relationships to understand sustainable leadership better.

Methodology: Exploring the Emerging Topic

Since sustainable leadership is a relatively new topic, most studies are in the exploratory phase. The majority adopts qualitative analysis or experiments to uncover its factors and dimensions. However, there is a need for more studies using qualitative methods like critical incident techniques and personal interviews. Additionally, future research should employ quantitative studies to develop scales and test them across various contexts, ensuring a more comprehensive understanding.

Discussions: Insights from the Bibliometric and TCCM analysis

This study conducted a bibliometric analysis and literature review using the TCCM framework on 24 articles about sustainable leadership and change management from the Scopus database. While not all these studies solely focus on leadership and change management, they contribute valuable insights. The initial bibliometric analysis reveals that the top countries in publishing sustainable leadership are the USA, Canada, Portugal, Australia, China, and the United Kingdom. Key journals include the Journal of Applied Behavioral Science and Business Ethics and Sustainability (Switzerland). The pivotal years in the publication landscape of this topic notably include 2017, 2021, and 2023, witnessing the highest number of published articles. These particular years stand out as peak scholarly activity, indicating a concentrated effort and heightened interest in advancing our understanding of the subject. The prominence of these years is not only a reflection of increased research output but also suggests significant developments or shifts in the discourse during these periods. As we delve into the specifics of these critical years, we can uncover the driving forces behind the surge in published articles, facilitating a more nuanced exploration of the topic's evolution and relevance over time.

The researchers used a special software called VOS Viewer for keyword co-occurrence analysis to find the essential words in the study of sustainable leadership. Since sustainable leadership is a new idea, the mentioned keywords are crucial. The top impactful keywords found were "organizational change," "leadership," "ethical leadership," "collaborative leadership," "human," "ethics," "transformational leadership," "sustainability," "sustainable development," and "sustainable leadership." The author also tried to understand the literature about sustainable leadership using the TCCM analysis. He wanted to know about the theories used, the context in which the studies were done, the characteristics studied, and the methods used. When looking at the types of theories used, they found that different leadership style theories, change management theories, corporate sustainability theories, social development theories, behavioral theories, and learning theories were used for the previous study. Only a few studies used theories for sustainable leadership. Most studies are still exploring, so they have yet to use or develop theories much. The context in which sustainable leadership is studied is in different sectors and performance areas. The characteristics mentioned in previous studies include Innovation, profitability, change management, happiness, and more, which come before sustainable leadership. As outcomes or results of sustainable leadership, they found stakeholders' profit, organizational performance, employee performance, reputation, satisfied workforce, and more. Figure 7.7.

In terms of methods used, most studies use surveys (both qualitative), case studies, and review studies. Only a few studies use quantitative or empirical studies to understand sustainable leadership better. Bibliometric analysis, which looks at how often articles are cited, shows that some topics are more popular or necessary because they get cited more. However, even if an article significantly impacts it, it might get few citations. Since sustainable leadership's impact on organizational change is new, recent papers may have few citations, but they could become more popular over time. That is why doing this kind of analysis every decade is essential to see how the field develops. The TCCM analysis is helpful for future researchers because it gives a quick overview of what theories are used, in what context, what characteristics are studied, and what methods are used in the studies of sustainable leadership. This method can help identify gaps and future directions for studying sustainable leadership.

Practical Implications of the Study

The study results enable policymakers to customize plans incorporating sustainable practices and promoting long-term beneficial transformation. Furthermore, the bibliometric analysis directs organizational policies and educational curricula toward paradigms in sustainable leadership and

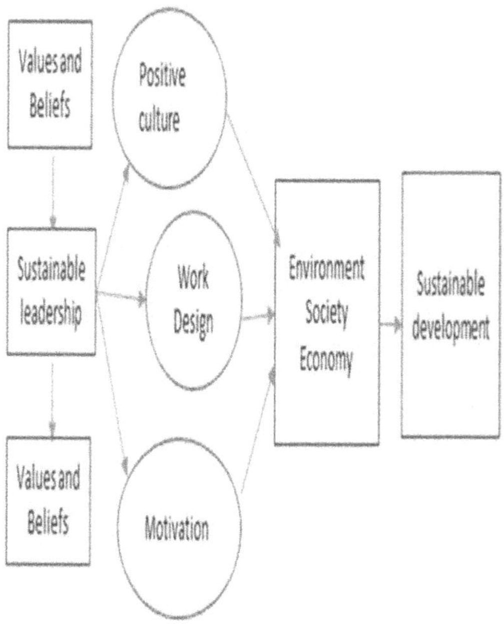

Figure 7.7 Conceptual map
Source: Authors Compailation

change management that are contemporary and pertinent for practitioners and scholars. In addition to guiding practical implementations in companies, this research advances the continuous development of sustainable leadership practices.

Future Research Directions

This study conducted a thorough assessment and bibliometric analysis using the TCCM framework, producing meaningful data on previous and ongoing research. The findings provide a starting point for additional study, allowing for focused investigation and improved models. Clarifying leadership characteristics, developing a flexible conceptual framework, and examining the causes and consequences of sustainable leadership are some of the major concerns. The study promotes confirmatory and empirical research, highlighting the necessity of more comprehensive bibliometric analyses to improve knowledge of sustainable leadership.

Conclusion

This study emphasizes the significance of sustainable leadership in impacting transformative processes within firms. With the help of a thorough bibliometric analysis review and the lens of the theory development, context, characteristics, and methodology (TCCM) framework, this study clarifies the critical role that sustainable leadership practices play in bringing about organizational change. Furthermore, in addition to simply highlighting noteworthy patterns and pinpointing areas of incomplete information, the results give insightful counsel to scholars and executives who aim to advance the field of sustainable organizational growth. This study adds to the current discussion on the effectiveness of leadership in creating long-lasting change. It provides a solid framework for future research and valuable information for practical application in organizational settings. It emphasizes how influential leaders are in embracing sustainable practices as catalysts for significant and long-lasting organizational transformation by synthesizing theoretical insights and valuable data.

References

Acton, K. S. (2021). School leaders as change agents: do principals have the tools they need? *Management in Education*, 35(1), 43–51.

Akcay Kasapoglu, O. (2018). Leadership and organization for the companies in the process of industry 4.0 transformation. *International Journal of Organizational Leadership*, 7, 300–308.

Asbari, M., Hidayat, D. D., and Purwanto, A. (2021). Managing employee performance: from leadership to readiness for change. *International Journal of Social and Management Studies*, 2(1), 74–85.

Attah, E. Y. (2017). Effective leadership and change management for sustainable. *Public Administration and Management Research*, 4(2), 37–42.

Bansal, P., and DesJardine, M. R. (2014). Business sustainability: it is about time. *Strategic Organization*, 12(1), 70–78.

Bolden, R. (2016). Leadership, management and organisational development. In Gower Handbook of Leadership and Management Development. Routledge, pp. 117–132.

Ferdig, M. A. (2020). Leadership for sustainability: generative engagement for change. *Encyclopedia of the World's Biomes*, 1:358–367.

Graetz, F. (2000). Strategic change leadership. *Management Decision*, 38(8), 550–564.

Guzmán, V. E., Muschard, B., Gerolamo, M., Kohl, H., and Rozenfeld, H. (2020). Characteristics and skills of leadership in the context of industry 4.0. *Procedia Manufacturing*, 43, 543–550.

Hajian, M., and Kashani, S. J. (2021). Evolution of the concept of sustainability. from brundtland report to sustainable development goals. In Sustainable Resource Management. Elsevier, pp. 1–24.

Hargreaves, A., and Fink, D. (2012). Sustainable Leadership. John Wiley & Sons.

Hunger, J. David (2020) Essentials of Strategic Management, 2, 542.

Iqbal, Q., Ahmad, N. H., Nasim, A., and Khan, S. A. R. (2020). A moderated-mediation analysis of psychological empowerment: sustainable leadership and sustainable performance. *Journal of Cleaner Production*, 262, 121429.

Jeronen, E. (2023). Economic sustainability. In Encyclopedia of Sustainable Management. Springer, pp. 1257–1263.

Kutlu, K. (2020). World Commission on Environment and Development (Brundtland Commission). The Palgrave Encyclopedia of Global Security Studies.

Leal Filho, W., Eustachio, J. H. P. P., Caldana, A. C. F., Will, M., Lange Salvia, A., Rampasso, I. S., et al. (2020). Sustainability leadership in higher education institutions: an overview of challenges. *Sustainability*, 12(9), 3761.

Leal Filho, W., Tripathi, S. K., Andrade Guerra, J. B. S. O. D., Giné-Garriga, R., Lovren, V. O., and Willats, J. (2019). Using the sustainable development goals towards a better understanding of sustainability challenges. *International Journal of Sustainable Development and World Ecology*, 26(2), 179–190.

Martín-Martín, A., Thelwall, M., Orduna-Malea, E., and López-Cózar, E. D. (2021). Google scholar, microsoft academic, scopus, dimensions, web of science, and openCitations' COCI: a multidisciplinary comparison of coverage via citations. *Scientometrics*, 126(1), 871–906.

Fyke, Jeremy P, and Patrice M Buzzanell. (2013). "The Ethics of Conscious Capitalism: Wicked Problems in Leading Change and Changing Leaders." *Human Relations* 66 (12): 1619–43.

Garrison, D Randy, and Norman D Vaughan. (2013). "Institutional Change and Leadership Associated with Blended Learning Innovation: Two Case Studies." *The Internet and Higher Education* 18: 24–28.

Metcalf, L., and Benn, S. (2013). Leadership for sustainability: an evolution of leadership ability. *Journal of Business Ethics*, 112, 369–384.

Metwally, Dina, Pablo Ruiz-Palomino, Mohamed Metwally, and Leire Gartzia. (2019). "How Ethical Leadership Shapes Employees' Readiness to Change: The Mediating Role of an Organizational Culture of Effectiveness." *Frontiers in Psychology* 10: 434635.

Muff, K., Liechti, A., and Dyllick, T. (2020). How to apply responsible leadership theory in practice: a competency tool to collaborate on the sustainable development goals. *Corporate Social Responsibility and Environmental Management*, 27(5), 2254–2274.

Myers, C., and Rushbrook, E. (2024). Leadership. In Disaster Health Management. Routledge, pp. 335–346.

Nandedkar, A., Mbindyo, M., O'Connor, R. J., and O'Connor, T. C. (2023). Academic advising leadership and change management. In Academic Advising Administration. Routledge, pp. 68–84.

Nguyen, H. T., Le, D. M. D., Ho, T. T. M., and Nguyen, P. M. (2021). Enhancing sustainability in the contemporary model of CSR: a case of fast fashion industry in developing countries. *Social Responsibility Journal*, 17(4), 578–591.

Oberer, B., and Erkollar, A. (2018). Leadership 4.0: digital leaders in the age of industry 4.0. *International Journal of Organizational Leadership*, 7(4), 404–412.

Oriade, A., Osinaike, A., Aduhene, K., and Wang, Y. (2021). Sustainability awareness, management practices and organisational culture in hotels: evidence from developing countries. *International Journal of Hospitality Management*, 92, 102699.

Paraschiv, Dorel Mihai, Estera Laura Nemoianu, Claudia Adriana Langă, and Tünde Szabó. (2012). "Eco-Innovation, Responsible Leadership and Organizational Change for Corporate Sustainability." *Amfiteatru Economic Journal* 14 (32): 404–19.

Piwowar-Sulej, K., and Iqbal, Q. (2023). Leadership styles and sustainable performance: a systematic literature review. *Journal of Cleaner Production*, 382, 134600.

Pranckutė, R. (2021). Web of science (WoS) and scopus: the titans of bibliographic information in today's academic world. *Publications*, 9(1), 12.

Rehman, S., Sami, A., Haroon, A., and Irfan, A. (2019). Impact of sustainable leadership practices on public sector organizations: a systematic review of past decade. *Journal of Public Value and Administrative Insight*, 2(3), 1–5. https://doi.org/10.31580/jpvai.v2i3.927.

Saha, V., Mani, V., and Goyal, P. (2020). Emerging trends in the literature of value co-creation: a bibliometric analysis. *Benchmarking: An International Journal*, 27(3), 981–1002.

Shaik, A. S., and Dhir, S. (2020). A meta-analytical review of factors affecting the strategic thinking of an organization. *Foresight*, 22(2), 144–177.

Sharif, Monica M, and Terri A Scandura. (2014). "Do Perceptions of Ethical Conduct Matter during Organizational Change? Ethical Leadership and Employee Involvement." *Journal of Business Ethics* 124: 185–96.

Silvestri, A., and Veltri, S. (2020). Exploring the relationships between corporate social responsibility, leadership, and sustainable entrepreneurship theories: a conceptual framework. *Corporate Social Responsibility and Environmental Management*, 27(2), 585–594.

Singh, V. K., Singh, P., Karmakar, M., Leta, J., and Mayr, P. (2021). The journal coverage of web of science, scopus and dimensions: a comparative analysis. *Scientometrics*, 126, 5113–5142.

Stead, J. G., and Stead, W. D. (2014). Sustainable Strategic Management. Routledge.

Tipu, S. A. A. (2022). Organizational change for environmental, social, and financial sustainability: a systematic literature review. *Review of Managerial Science*, 16(6), 1697–1742.

Zhu, J., and Liu, W. (2020). A tale of two databases: the use of web of science and scopus in academic papers. *Scientometrics*, 123(1), 321–335.

8 Strategic research and development investments: analysing financial success in the drugs and chemicals industry of India

Ashutosh Parhi[1,a] *and Rabi Narayan Subudhi*[2,b]

[1]Research Scholar, School of Management, KIIT Deemed to be University, Bhubaneswar, India

[2]Senior Professor, School of Management, KIIT Deemed to be University, Bhubaneswar, India

Abstract

This empirical research investigates the influence of research and development (R&D)spending on the enterprise's financial success within the Indian pharmacological sector. The pharmaceutical industry faces rapid technological evolution and intense competition. In response to these challenges, strategic R&D investments have emerged as imperative for scientific progress and a fundamental strategy for gaining a competitive advantage. The study focuses on the period 2015 to 2023, examining 50 pharma enterprises listed on the national stock exchange (NSE) of the drugs and chemicals sector. Two key financial performance metrics- returns on assets (ROA), and earnings per share (EPS), are employed, providing a dual perspective on operational efficiency and market perception. The study incorporates control variables, including firm size, leverage ratio, liquidity ratio, and operating margin, to isolate the influence of R&D expenses on an enterprise's financial outcomes. The research contributes to a clear understanding of the complex dynamics of financial performance in the pharmaceutical sector. Practical implications for businesses, investors, and policymakers are highlighted while acknowledging limitations such as data availability constraints and industry-specific focus. The study sets the stage for future research directions, advocating for extended analysis durations, industry-specific inquiries, and the exploration of emerging technologies' impact on R&D.

Keywords: Earnings per share, financial performance, innovation, patent, pharmaceutical company, research and development, expenditure, returns on assets

Introduction

The pharmaceutical sector stands at the forefront of modern healthcare. It plays a pivotal role in addressing global health challenges and improving the quality of life for millions. In an era marked by rapid technological evolution and increasing complexities in disease patterns, the significance of robust research and development (R&D) in the pharmaceutical industry cannot be overstated (Nath et al., 2010). Pharmaceutical R&D serves as the driving force behind the discovery and development of innovative drugs, therapies, and medical technologies (Henard and McFadyen, 2012). The relentless pursuit of scientific breakthroughs in pharmaceutical research facilitates the creation of novel treatments and fostering innovation in healthcare solutions (Ramirez and Girdauskiene, 2013). As the healthcare situation undergoes rapid transformations and breakthroughs, the pharmaceutical sector is compelled to navigate a complex and dynamic environment. In this context, strategic R&D investments emerge not only as a response to the imperative of scientific progress but also as a fundamental strategy for securing a competitive advantage (Hsu

[a]ashutoshparhi200@gmail.com, [b]rabisubudhi@gmail.com

DOI: 10.1201/9781003564447-8

et al., 2013). The global pharmaceutical arena is witnessing heightened competition, with companies competing for market share, and, the trust of healthcare practitioners and consumers. In an environment characterized by heightened competition, the function of R&D initiatives serves as a critical determinant for organizational differentiation (Insee and Suttipun, 2023).

Pharmaceutical companies that actively invest in R&D tend to draw investors' attention because of their promising prospects for revenue growth, innovation, and long-term sustainability (Collins et al., 2011). A company that actively pursues R&D can effectively mitigate the risks associated with patent expirations and generate a strong competitive edge in the market by demonstrating its commitment to discovering and developing novel medications (Bae et al., 2012). The diversification of product portfolios resulting from successful R&D efforts contributes significantly to sustained revenue growth and market exclusivity, thereby augmenting the overall profitability of pharmaceutical companies (Kim et al., 2021). Furthermore, investors acknowledge that a proactive stance toward R&D not only fosters positive market perceptions but also positions companies as leaders within the industry, thereby instilling long-term investor confidence (Paul et al., 2010). In addition, pharmaceutical firms engaged in cutting-edge R&D emerge as attractive investments for those seeking financial returns and active participation in the dynamic landscape of healthcare innovation (Aslan et al., 2010).

The Indian pharmaceutical sector plays a crucial role in meeting 95% of the nation's healthcare demands while making substantial international contributions (Abrol, 2014). The anticipated trajectory for the market valuation of the Indian chemical industry involves a projected cost of USD 173 billion (2019) to USD 304 billion (2025), characterized by a Compound Yearly Growth Rate (CAGR) of around 9.3%. Within this sector, an annual growth rate of 9% is anticipated, contributing significantly to the Indian Gross Domestic Product (GDP) with a projected value of USD 300 billion. By the year 2025, the segment of specialty chemicals, constituting 22% of the industry, is poised to attain a valuation of USD 64 billion. By 2025, India is predicted to have a 5.5% worldwide share in specialized chemicals, up from 3% in 2015. The turning point for the industry traces back to the enactment of the Indian Patent Act in 1970. This legislative move, distinguishing between product and process patents, marked a departure from the industry's sluggish growth and fostered an environment conducive to expansion through regulatory leniency (Dasgupta and Stiglitz, 1980; Chaudhuri, 2007). This regulatory relaxation opened doors for diverse firms, democratizing industry participation that was once constrained by stringent norms. This shift induced heightened competition, compelling major players, particularly large pharmaceutical firms, to excel in producing high-quality drugs at competitive prices (Government of India Report, 2008). Beyond fostering entrepreneurial activity, this competitive landscape triggered a transformation in strategic behaviour. The dynamism observed in the market, characterized by intensified competitiveness, underscores the evolving nature of competition. This evolution places considerable emphasis not only on the interplay between costs and prices but also on the organizational agility of pharmaceutical firms to adapt swiftly to emerging market conditions and innovations (Gelei, 2003).

The pharmaceutical sector was chosen as the sample study of this research due to its pivotal part in addressing worldwide health challenges and improving the quality of life for millions worldwide. Robust R&D is essential in the pharmaceutical industry due to rapid technological evolution and increasingly complex disease patterns. Strategic R&D investments are critical not only for scientific advancement but also for securing a competitive edge in the pharmaceutical market. Understanding the dynamics of R&D investments in the pharmaceutical sector holds strategic implications for policymakers, industry stakeholders, and investors, informing discussions surrounding innovation policies, industry regulations, and economic development strategies. The Indian Patent Act of 1970, has fundamentally shaped the trajectory of the Indian pharmaceutical industry by fostering competition and innovation (Bhattacharjea and Sindhwani, 2014). Thus, analyzing their impact on R&D investments and competitiveness becomes an unexplored area of research.

Review of Literature

Research has long emphasized the pivotal role of R&D in shaping a corporation's future productivity, sales growth, and overall competitiveness (Morbey, 1988). Given the rapid pace of technological advancements in the global business landscape, leading firms must allocate resources toward R&D to fortify or enhance their market competitiveness (Shen et al., 2017). While numerous studies have studied the effect of R&D investment on enterprise performance, particularly in developed or industrialized nations, there remains a noticeable gap in the literature regarding R&D investment in developing countries. Some studies have partially explored the influence of R&D on firm performance using either accounting-based or market-based measures. However, none have comprehensively examined both aspects within the context of a developing country such as India. Hence, this paper is structured to address following two key research questions:

i. How does an R&D venture affect the enterprise performance of the pharmaceutical sector in developing countries like India, as measured by accounting-based metrics, ROA?
ii. How does an R&D venture affect the enterprise performance of the pharmaceutical sector in developing countries like India, as measured by market-based metrics, EPS?

To address these research questions, two research objectives are derived:

- To analyze the impact of R&D investment on the return on assets (ROA) of pharmaceutical companies, focusing on the Indian context.
- To examine the relationship between R&D investment and earnings per share (EPS) of pharmaceutical firms in India, a measure of firm performance.

Based on earlier research, Erickson and Jacobson (1992) discovered that combining expenditures of R&D and advertising, results in a comparative advantage, which raises profits per share (EPS) for businesses. Rao et al. (2013) employed empirical and comparative methodologies to investigate the impact of technical innovation in China and Japan. Their investigation focused on exploring the overall performance of corporations through the amount spent on research and development activities. The importance of intangible assets and intellectual capital has been highlighted in several studies; according to Bhasin (2016), these assets account for around 87% of the market value for companies in developed markets and roughly 47% for listed companies in emerging nations (Anghel, 2008). Feinberg and Majumdar (2001) noted that multinational corporations (MNCs) were the only ones in the Indian pharmaceutical business to spend significantly on research and development, which forced Indian enterprises to step up their efforts to keep up with their international competitors. R&D expenditure and future financial returns are positively correlated, according to Cincera et al. (2009). According to Ehie and Olibe (2010), industry concentration, leverage, company size, and R&D spending all have a significant impact on a company's financial performance. R&D expenditure possess a significant impact on business success, according to Freihat et al. (2017), which is consistent with findings from developed nations. Their study revealed that spending in R&D each year yields future advantages, including heightened market presence, enhanced stock value, and an improved corporate image. According to Anagnostopoulou (2008), R&D and future market performance are positively correlated. Additionally, VanderPal (2015) discovers a substantial positive relationship between the company's market value, profitability, and R&D. Su et al. (2020) conducted an empirical investigation into the correlation between the extent of R&D expense and the operative performance of Chinese SMEs. Their findings indicate that, upon surpassing a threshold of 5% in R&D investment, the association between R&D expense and the enterprise's operative performance results exhibits an adverse trend for small and medium-sized enterprises. The future profits per share of a company are positively correlated with its E&D expense for the Indian automobile industry (Mondal and Sarkar, 2020). As per Ballester et al. (2003) and Monahan (2005), organizations that allocate greater resources to research and development (R&D) are likely to generate increased future earnings. As per Connolly and Hirschey (2005), the enterprise size influences both the amount of research and development activity invested and the quality of the results produced by that expenditure. The effect of financial leverage on R&D expenditure is demonstrated by another study. Businesses with greater debt tend to spend less on research and development (Nam et al., 2003). R&D spending has an impact on firm value, as evidenced by data from South Korea that shows a positive connection between the stock price and the company's R&D spending (Hana, 2004).

Consequently, based on this discussion, the present study postulates the following hypotheses:

(H1): A positive correlation exists between R&D and the financial performance of enterprises, employed by (EPS).

Konak et al., (2016), in their study, elucidate the influence of Research and Development activities on corporate financial performance, proxy by ROA. The outcomes of their investigation unveiled an adverse influence of R&D doings on the enterprise's financial success. Similarly, Ayaydin et al., (2014) aimed to identify determinants of financial performance, observing the detrimental effect of R&D initiatives on an enterprise's financial performance. Cui and Mak (2002) reported a negative correlation between (ROA) and a positive association between R&D activities and firm value (Tobin's Q). Jannine (2008) posited that the use of technology negatively impacts the profitability of firms by hindering their R&D activities and affecting the value of companies negatively. Ahmed et al. (2011) observed that the level of research and development intensity is adversely associated with an enterprise's market value.

Contrarily, Bouaziz, (2016) concluded that no relation exists between R&D expense and the enterprise's performance. Jian and Zhenji (2016) draw a similar conclusion, that there is no significant relationship between the enterprise's productivity, and R&D investment. Conversely, Yanni Wang et al. (2016) hypothesized that businesses investing in research and development tend to be more valuable to the market, suggesting that businesses should formulate policies supporting long-term growth. In contrast, Serap and Kenan (2016) observed a negligible correlation between research and development spending and gross profit, while Selvarajah and Sheena (2017) discovered an adverse link between R&D and the business success of IT firms. Consequently, drawing insights from these studies, the present research formulates the following hypotheses to study it in Indian context:

(H2): *A negative association is present between R&D and the financial performance of enterprises, as measured by (ROA).*

An evident gap in contemporary academic research pertains to the insufficient exploration of R&D initiatives on both the accounting and market-based dimensions of an enterprise's financial performance. Previous studies have tended to focus on either accounting metrics or market-based indicators in isolation, thus leaving a discernible research gap. The current empirical study endeavors to address this gap through the adoption of a holistic approach, recognizing the multifaceted nature of financial performance assessment by concurrently considering both accounting and market-based measures. The primary objective of this research is to offer a nuanced understanding of the intricate relationship between R&D operations and the overall financial health of a company. Additionally, the study emphasizes elucidating the strategic value embedded in R&D expenditures as a means of attaining a competitive advantage within the corporate landscape. By undertaking such an approach, this research aspires to provide clarity on how R&D endeavors contribute to shaping a corporation's financial performance.

Data and Research Methodology

Sample

This research explores the relationship between R&D expenditures and the enterprise's financial performance listed on the national stock exchange (NSE) under the Drugs and Chemicals Sector. The study spans the time frame of 2015 to 2023. Drawing data from the center for monitoring of Indian economy (CMIE) Prowess IQ database, the study encompasses the entire set of entities listed in the drugs and chemicals index, totaling 145 firms. To ensure robust analysis, companies with consecutive missing data for two years were systematically excluded from the study. Consequently, the final dataset comprises a sample of 50 firms, yielding a total of 450 observations.

Dependent Variable

The response variable of the study is the financial performance of the enterprises within the Indian pharmaceutical sector. The assessment of the financial efficiency of these entities is conducted through the utilization of two discrete metrics. The first metric employed is ROA, an accounting-based measure designed to evaluate profitability. The second metric utilized is EPS, a market-based measure that provides insights into both shareholders' return and the valuation of the firm. These chosen metrics collectively serve as quantitative indicators for the financial health and performance of the examined pharmaceutical companies, enabling a comprehensive evaluation grounded in both accounting and market-oriented perspectives. This dual approach allows for a thorough analysis of the connection between research and development investment and financial success, encompassing both operational efficiency and market perception.

Independent Variable

The R&D expenditure is the explanatory variable of the study. R&D expenditure encompasses investments directed towards advancing products or services, as well as addressing existing or prospective challenges. During data collection, we encountered discrepancies in the disclosure of research and development investment, with some companies explicitly reporting it as "R&D expenditure". To address this inconsistency and ensure comprehensive analysis, R&D expenditure was calculated for such firms using a standardized approach of 0.05% of total revenue earned. This ensures a consistent and accurate representation of R&D investment across all firms within the study sample.

Control Variables

The current study utilizes several control variables that could potentially influence the observed results, including firm size (log value of total assets), leverage ratio (long-term debt to total assets), liquidity ratio (current assets to current liabilities), and operating margin (EBITDA). These variables account for important characteristics that might affect a corporation's financial performance independent of R&D expenditure, allowing for a more accurate valuation of the influence of R&D on an enterprise's financial outcomes.

Large corporations generally have greater access to resources and economies of scale, potentially leading to higher financial performance. Therefore, controlling company size helps isolate the specific influence of research and development spending on performance. The extent of debt financing (leverage) can influence a company's risk profile and financial performance. Controlling the leverage ratio ensures that observed variations in financial performance are primarily driven by R&D and not solely due to different debt levels. A firm's ability to fulfill immediate financial obligations (liquidity) can impact its financial health and stability. Controlling liquidity ratio helps isolate the long-term effect of R&D expenditure on performance and avoid confounding results from short-term liquidity issues. Operating margin reflects a firm's efficiency in generating profit from its core operations. Controlling the operating margin helps isolate the unique influence of R&D spending on a firm's financial success, hence differentiating it from the general profitability of the firm's core business.

Model Specification and Research Methodology

This study undertakes an examination of the influence of R&D initiatives on the financial performance of the listed Indian pharmaceutical corporations. Employing panel regression analysis with a balanced dataset, the research scrutinizes the association of R&D expenditures and two pivotal financial success metrics: ROA and EPS. The chosen analytical approach seeks to meticulously isolate the distinct impact attributable to R&D expenditures on financial performance, concurrently accounting for other pertinent factors that may exert influence. By adopting this methodological framework, the study seeks to provide a comprehensive assessment of the impact of research and development initiatives on the financial outcomes of the drug and chemicals sector firms. The empirical models outlined for analysis are as follows:

$$ROA = \alpha + \beta 1RDE + \beta 2FS + \beta 3LEV + \beta 4LQ + \beta 5OM \in \quad (1)$$

$$EPS = \alpha + \beta 1RDE + \beta 2FS + \beta 3LEV + \beta 4LQ + \beta 5OM + \in \quad (2)$$

Where:

ROA: Firm profitability; return on assets (PAT/total assets)

EPS: Firm market performance; earning per share (EPS)

RDE: Research and development expenditure

FS: Firm size (natural logarithm value of total assets)

LEV: Debt ratio (non-current debt/total assets)

LQ: Liquidity ratio (current assets/current liabilities)

OM: Operating margin (earnings before interest, tax, depreciation, and amortization)

This study employs a multi-dimensional approach to analyze the influence of R&D expenditure on the financial outcomes of the listed Indian pharmaceutical firms. Descriptive statistics provide an initial overview of the data, while simple regression (OLS) and Pearson correlation analysis delve deeper into the associations between variables. Additionally, diagnostic tests such as normality, variance inflation factor (VIF), and heteroskedasticity tests assess the suitability of the balanced panel data for further analysis. Two-panel regression models are then estimated: fixed effects model and random effects model. To choose the most appropriate model, the Hausman specification test is conducted, ensuring a robust and valid assessment of the employed models.

Data Analysis and Results

Summary Statistics

The given Table 8.1 provides a comprehensive overview of key financial variables, offering valuable insights into the financial performance of the entities under study. The ROA is a critical metric, indicating the efficiency of asset utilization. The mean ROA of 7.49% suggests a positive average return, but the significant standard deviation of 8.41% underscores the variation in performance across entities. EPS reflects profitability, with a mean of 25.05. The substantial standard deviation of 51.46 highlights considerable variability, signifying diverse financial performances. Research and development expenses (RDE) exhibit a wide range, with a mean of 1853.09, indicating diverse investment patterns in innovation. firm size (FS), measured with a mean of 4.30, showcases the overall stability and robustness of the entities, while leverage (LEV) at a mean of 0.148 provides insights into the capital structure diversity. Liquidity (LQ) at a mean of 2.08 indicates the ease with which assets can be converted to cash, and operating margin (OM) at a mean of 0.119 reflects operational efficiency. The significance lies in these metrics serving as key indicators for stakeholders, enabling a clear understanding of financial performance, risk, and strategic decision-making for the entities.

Co-relation Analysis and Variance Inflation Factor

Table 8.2 presents a thorough analysis of the interrelationships among key financial variables, utilizing both correlation coefficients and variance inflation factor (VIF) values. The moderate positive correlation of 0.5167 between the ROA and the EPS indicates that companies with higher ROA tend to exhibit elevated EPS, underscoring the positive influence of profitability on earnings distributed to shareholders. A weak positive correlation of 0.0556 between

Table 8.1 Descriptive statistics outcomes.

Variable(s)	Obs.	Mean Value	Std. dev	Min Value	Max Value
ROA	450	7.49	8.41	-50.16	81.17
EPS	450	25.05	51.46	-378.6	440.34
RDE	450	1853.09	3545.06	1.2	23101.3
FS	450	4.30	.624	3.107	5.665
LEV	450	.148	.155	.0034	.961
LQ	450	2.08	1.21	.3322	7.894
OM	450	.119	.163	-1.39	.718

Source: Estimated by the authors

Table 8.2 Co-relation and VIF.

Variables	ROA	EPS	R&D	Company size	Debt ratio	Liquidity ratio	EBITDA	VIF
ROA	1							-
EPS	0.516	1						-
R&D	0.055	0.145	1					2.15
Company size	0.080	0.230	0.694	1				2.08
debt ratio	-0.479	-0.191	-0.157	0.007	1			1.65
Liquidity ratio	0.465	0.203	0.070	0.102	-0.347	1		1.50
EBITDA	0.852	0.440	0.142	0.208	-0.500	0.472	1	1.32
								1.74

Note: Variance inflation test (VIF) is conducted for the multicollinearity issues

Source: Estimated by the authors

ROA and R&D expenditure suggests a limited linear relationship, while the 0.0805 correlation with Company Size implies a modest association between ROA and the size of the company. The moderate negative correlation of –0.4794 between ROA and Debt Ratio indicates that higher debt levels may be associated with lower ROA, reflecting a potential negative impact on profitability. Moreover, the moderate positive correlation of 0.4658 between ROA and Liquidity Ratio implies that firms with elevated liquidity levels may demonstrate higher ROA, underscoring the favorable association between liquidity and profitability. The robust positive correlation of 0.8524 between ROA and earnings before interest, taxes, depreciation, and amortization (EBITDA) indicates a significant relationship, where increased EBITDA is linked to higher ROA.

The moderate positive association of 0.5167 between EPS and ROA implies a significant association, indicating that profitability positively influences earnings distributed to shareholders. The positive correlation of 0.1450 with R&D expenditure suggests a moderate relationship, indicating that companies investing more in R&D may experience a positive impact on their earnings per share. The 0.2308 positive correlation with company size implies a moderate association, suggesting that larger companies may have higher EPS. Conversely, the negative correlation of –0.1912 between EPS and debt ratio indicates that higher levels of debt are associated with lower EPS, while the positive correlation of 0.2033 with liquidity ratio suggests that companies with higher liquidity ratios may experience higher EPS. The significant relationship of 0.44 between EPS and EBITDA underscores that overall earnings performance positively influences EPS.

The variance inflation factor (VIF) values presented in Table 8.2 serve as a critical diagnostic tool for assessing multicollinearity among independent variables, a phenomenon where variables are highly correlated and may introduce challenges in regression analysis. The variance Inflation Factor quantifies the extent to which the variance of an independent variable is inflated due to correlations with other independent variables in the model. In the context of this analysis, the VIF values for each independent variable—R&D, company size, debt ratio, liquidity ratio, and EBITDA—are all found to be below the conventional threshold of 10. This outcome signifies a low risk of multicollinearity, indicating that the independent variables are not excessively correlated.

Regression Results

In the regression analysis (Table 8.3), two separate models were implemented to explore the connection between the independent variable (R&D) and financial performance indicators (ROA and EPS). The first model explored the impact of R&D expenditure, company size, debt ratio, liquidity ratio, and EBITDA on ROA, with coefficients revealing their influences. For instance, R&D expenditure showed a coefficient of –0.0000274, while company size displayed a coefficient of –1.196523. Notably, the liquidity ratio exhibited a positive impact on ROA with a coefficient of 0.5209287, and EBITDA contributed significantly with a coefficient of 41.76667. In the second model, the analysis shifted to Earnings Per Share (EPS) as the dependent variable. Here, R&D expenditure, company size, debt ratio, liquidity ratio, and EBITDA similarly showcased their respective impacts. For instance, company size had a substantial positive effect on EPS with a coefficient of 13.20577, while debt ratio exhibited a positive influence with a coefficient of 3.924066. EBITDA emerged as a pivotal factor, displaying a considerable positive impact on EPS with a coefficient of 131.5953. Both models were based on 450 observations, and the F-statistics of 257.06 for ROA and 24.32 for EPS indicated the overall significance of the models. The R-squared values of 0.7432 for ROA and 0.2150 for EPS signified the proportion of variance explained by the independent variables, while the adjusted R-squared values of 0.7404 for ROA and 0.2061 for EPS, accounting for the number of predictors, underscored the models' goodness-of-fit. These outcomes add to a deeper knowledge of the complex relationships between the analyzed independent variables and their respective impacts on financial performance indicators, namely ROA and EPS.

Panel Regression Outcome

Table 8.4 shows the outcomes of a panel regression analysis employing a fixed effects (FE) model, elucidating the relationships between firm performance, proxied by ROA and EPS. The adoption of the FE model is particularly salient, given its capacity

Table 8.3 Regression result (OLS).

	Dependent variable			
	ROA		EPS	
Independent variables	Coef. value	Std. error	Coef. value	Std. error
R&D expenditure	-.0000274 (0.740)	.0000823	-.000347 (0.694)	.0008805
Company size	-1.196523***(0.012)	.4746869	13.20577*** (0.010)	5.076436
Debt ratio	-2.536436 (0.111)	1.589043	3.924066 (0.817)	16.99367
Liquidity ratio	.5209287 (0.007)	.1913049	-.2210869 (0.914)	2.045869
EBITDA	41.76667*** (0.000)	1.583739	131.5953*** (0.000)	16.93696
Observations	450		450	
F (5, 444)	257.06		24.32	
R square	0.7432		0.2150	
Adj. R square	0.7404		0.2061	

The "p" values are displayed in parentheses. $**p < .05$, $***p < .001$, and $*p < .10$

Source: The authors' estimation

Table 8.4 Panel regression result (FE Model).

	Dependent Variable (Firm Performance)			
	ROA		EPS	
Independent variable	Coef. Values	Std. error	Coef. Values	Std. error
R&D expenditure	-.0003693** (0.046)	.0001847	-.0007564* (0.639)	.0016101
Company size	-4.133255*** (0.002)	1.326785	35.84023*** (0.002)	11.56881
Debt ratio	-14.79958*** (0.000)	2.817417	-6.253831 (0.799)	24.56626
Liquidity ratio	.8276364*** (0.001)	.2513001	-4.279516** (0.052)	2.191193
EBITDA	47.01927*** (0.000)	1.765108	146.0302*** (0.000)	15.39073
R^2	0.6665		0.1882	

The "p" values are displayed in parentheses. $^{**}p < .05$, $^{***}p < .001$, and $^{*}p < .10$

Source: The authors' estimation

to control for time-invariant unobserved heterogeneity at the firm level, thereby ameliorating potential biases arising from omitted variables. This methodological choice ensures a robust examination of the impact of R&D expenditure, company size, debt ratio, liquidity ratio, and EBITDA on firm performance. The Breusch-Pagan LM test is used to determine whether the data are heteroskedastic before conducting an empirical data analysis. Heteroskedasticity was indicated by the test findings (χ^2 = 40.05 and p-value =.000). The Hausman test is used to determine whether fixed effect (FE) or random effect (RE) models are appropriate for panel regression. The FE and RE models' estimated coefficients are compared using the Hausman test (Hausman, 1978). The RE model's effectiveness and consistency serve as the test's null hypothesis. As opposed to this, the alternate hypothesis assumed that the FE regression model would be reliable and operative. The Husman test's outcome (χ^2 = 3.20 and the p = .00) indicates the FE model is appropriate for additional research since the null hypothesis is rejected.

the statistically significant coefficient for ROA (p = 0.046) signifies that R&D expenditures exert a discernible influence on a firm's return on assets. Conversely, the non-significant coefficient for EPS (p = 0.639) implies a lack of statistical significance for the relationship with earnings per share. The highly significant coefficients for both ROA and EPS (p = 0.002) underscore the substantial impact of company size on firm performance, underscoring its pivotal role as a predictor. Regarding the debt ratio, the highly significant negative coefficient for ROA (p = 0.000) indicates an inverse association with returns on assets. However, the non-significant coefficient for EPS (p = 0.799) suggests that the debt ratio may not significantly influence earnings per share. The highly significant coefficient for ROA (p = 0.001) indicates a positive relationship between liquidity and return on assets. The marginally significant coefficient for EPS (p = 0.052) suggests a potential impact of liquidity on earnings per share at the 5% significance level. Notably, the highly significant coefficients for both ROA and EPS (p = 0.000) affirm the substantial influence of EBITDA on firm performance, signifying that higher earnings before interest, taxes, depreciation, and amortization are associated with superior return on assets and increased earnings per share. The reported R-squared values denote the proportion of variance in firm performance explained by the independent variables, attesting to the overall goodness-of-fit of the model.

The observed negative association between R&D expenditure and enterprise performance aligns with the notion that R&D expenditures may initially lead to decreased profitability in the short run. This can be attributed to the high costs associated with R&D activities without immediate returns. However, it's important to reflect the possible long-term aids of R&D expenses in terms of innovation, product development, and market competitiveness, which may not be fully captured in the short-term analysis provided by the panel regression model. The negative relationship can be justified based on theoretical considerations and industry norms. Despite the short-term negative relationship observed in the regression analysis, increasing R&D expenditure may still be warranted for several reasons. Firstly, R&D is crucial for driving innovation and maintaining competitiveness in the pharmaceutical sector, where technological advancements and product differentiation are key drivers of success. Secondly, while the immediate financial returns may not be evident, investing in R&D can position the firm for long-term growth and sustainability by introducing new products, enhancing capabilities, and adapting to changing market demands. Therefore, while the empirical findings may indicate short-term challenges associated with R&D expenditure, strategic considerations, and industry expectations may still justify its continued investment.

In summation, these findings, derived from the FE model and p-values, augment our comprehension of the intricate relationships within the financial landscape, providing valuable insights for practitioners and policymakers to inform decision-making.

Discussion

The present investigation examines the effect of R&D spending on the financial performance of enterprises in the Indian drugs and chemicals sector. Table 8.2 presents the VIF and correlation coefficients to illustrate the interrelations among variables and address multicollinearity concerns. Notably, a moderate positive correlation of 0.4658 between ROA and EPS emphasizes the positive impact of profitability on shareholder earnings. The weak to moderate correlations with R&D expenditure, company size, debt ratio, and liquidity ratio provide insights into their relationships with ROA. The substantial correlation of 0.8524 between ROA and EBITDA signifies a significant relationship.

Table 8.3 presents regression results for both ROA and EPS models. Company size and EBITDA emerge as significant predictors for both financial performance indicators. Debt ratio demonstrates a negative effect on ROA, highlighting the importance of managing debt levels. Liquidity ratio positively influences both ROA and EPS. These regression models offer a comprehensive framework for understanding key variables' influence on financial performance.

Table 8.4 employs a fixed effects (FE) model in a panel regression analysis, providing insights while controlling for time-invariant unobserved heterogeneity. Results highlight the substantial impact of EBITDA on both ROA and EPS, with company size remaining a significant predictor. Debt ratio exhibits a negative association with ROA. The nuanced relationships unveiled by the FE model contribute valuable information for informed decision-making. These analyses contribute to a nuanced understanding of multifaceted relationships within the financial landscape, guiding strategic decisions for practitioners and policymakers to enhance financial performance.

In the regression analysis presented in Table 8.3, the affiliation between ROA and R&D spending is scrutinized. The p-value is 0.740, exceeding the conventional significance level of 0.05. Consequently, based on this p-value, the null hypothesis of no significant relationship between ROA and R&D expenditure is not rejected. Thus, the evidence does not adequately support claiming a statistically significant association between ROA and R&D expenditure. Similarly, the affiliation between EPS and R&D spending is examined, with a p-value of 0.694 above the significance level of 0.05, suggesting an insignificant relationship between EPS and R&D expenditure.

The panel regression analysis, employing a FE model in Table 8.4, aims to solve intricate relationships between critical financial performance metrics (ROA and EPS) and the variable of interest, R&D expenditure. Coefficients and their corresponding p-values are essential for the examination. The correlation between a firm's return on assets and its R&D expenditure is statistically significant and negative, indicating a clear effect of R&D spending on ROA. Conversely, the p-value of 0.639 in the relationship between EPS and R&D expenditure within the FE model does not provide adequate support to reject the null hypothesis, indicating the absence of a statistically significant relationship.

Conclusion

It is found that research and development (R&D) expenditure, company size, debt ratio, liquidity ratio, and earnings before interest, taxes, depreciation, and amortization (EBITDA) significantly influence Return on assets (ROA). Specifically, R&D expenditure shows a negative effect on firm performance in the short run, while company size and debt ratio have consistently negative effects on ROA. Conversely, liquidity ratio and EBITDA exhibit positive relationships with the financial performance of the firm. Notably, it's observed that while R&D expenditure initially decreases ROA, there's a potential for long-term profitability through innovative practices and patent applications. These findings underscore the complex interplay of financial and operational factors in determining a company's performance, suggesting the need for strategic management approaches to balance short-term profitability with long-term innovation and growth objectives.

This research paper offers an in-depth examination of the pharmaceutical sector, explaining the effect of R&D spending on shaping the financial success of enterprises. This report, which focuses on the Indian pharmaceutical sector, not only documents the industry's transformational path but also explains the strategic relevance of R&D investment. In a rapidly evolving environment marked by technological advancements and intense competition, pharmaceutical enterprises that actively invest in R&D are better positioned to secure a competitive advantage.

Contribution

By focusing on the Indian pharmaceutical sector, the paper offers insights into industry-specific dynamics and the strategic behavior of pharmaceutical firms regarding R&D investments. It identifies significant predictors of financial performance and highlights the nuanced relationships between R&D spending, profitability, and firm value. The paper discusses the policy implications of its findings, suggesting that policymakers can formulate laws to encourage R&D expenditure and boost economic expansion in the pharmaceutical sector. This empirical evidence can inform policy discussions surrounding innovation policies and industry regulations. Thus, the research paper contributes to both theoretical understanding and empirical knowledge regarding the impact of R&D spending on financial performance in the Indian pharmaceutical sector. It links the gap between academic research and real-world applications, contributing valuable insights for stakeholders including businesses, investors, policymakers, and researchers.

Limitations

The research is subject to several limitations. The unavailability of persistent data throughout the study led to a 9-year timeframe, potentially limiting the capture of long-term trends. The sample size was significantly reduced from 145 to 50 companies due to missing data, introducing potential sample selection bias. The exclusive focus on the pharmaceutical sector restricts the generalizability of findings to other industries, and the confinement to the Indian context limits global applicability.

Future Research Scope
Future research can extend analysis durations, delve into industry-specific inquiries, and utilize comparative assessments alongside global benchmarks. This approach is crucial for obtaining a nuanced comprehension of the relationship between a company's investment in R&D and its financial performance. Incorporating technological innovation metrics, exploring non-linear relationships, and employing qualitative research methodologies such as expert interviews would enhance comprehensiveness. Cross-sectional analyses, external factor considerations, and policy implications assessments are recommended for enriched studies. Future research could explore the long-term impact of emerging technologies on pharmaceutical R&D, investigate global collaboration dynamics, and contribute valuable insights for industry resilience and global health challenges.

References

Abrol, D. (2014). Technological Upgrading, Manufacturing, and Innovation: Lessons from Indian Pharmaceuticals. Institute for Studies in Industrial Development.

Ahmed, Kamran, John Hillier, and Elisabeth Tanusasmita. "R&D profitability, intensity and market-to-book: evidence from Australia." *Accounting Research Journal* 24, no. 2 (2011): 150–177.

Anagnostopoulou, S. C. (2008). R&D expenses and firm valuation: a literature review. *International Journal of Accounting and Information Management*, 16(1), 5–24.

Anghel, I. (2008). Intellectual capital and intangible assets analysis and valuation. *Theoretical and Applied Economics*, 3(3), 75.

Aslan, A., Kula, F., and Kaplan, M. (2010). New evidence on the persistence of profit in Turkey with first and second generation panel unit root tests. METU Studies in Development, 37, 25–40.

Ayaydın, Hasan, and İbrahim Karaaslan. "The Effect of Research and Development Investment on Firms'financial Performance: Evidence From Manufacturing Firms In Turkey." *Bilgi ekonomisi ve yönetimi dergisi* 9, no. 1 (2014): 23–39.

Bae, K.-H., Ozoguz, A., Tan, H., and Wirjanto, T. S. (2012). Do foreigners facilitate information transmission in emerging markets?. *Journal of Financial Economics*, 105(1), 209–227.

Ballester, M., Garcia-Ayuso, M., and Livnat. J. (2003). The economic value of the R&D intangible asset. *European Accounting Review*, 12(4), 605–633.

Bhasin, M. L. (2016). Intellectual capital measurement and disclosure: a new 'Paradigm' in financial reporting. *The Journal of Economics, Marketing and Management*, 4(4), 1–16.

Bhattacharjea, A., and Sindhwani, F. (2014). Competition issues in the Indian pharmaceuticals sector. *Center for Development Economics, Delhi School of Economics. New Delhi: CUTS-CIRC*, 1–92.

Bouaziz, Zied. "The impact of R&D expenses on firm performance: Empirical witness from the Bist Technology Index." *Journal of Business Theory and Practice* 4, no. 1 (2016): 51–60.

Chaudhuri, S. (2007). Is product patent protection necessary in developing countries for innovation? R&D by Indian pharmaceutical companies after TRIPS. *Agenda*, 9(12h00).

Cincera, M., Ortega-Argilés, R., and Moncada-Paternò-Castello, P. (2009). The Performance of top R&D Investing Companies in the Stock Market. No. 14/2009. IPTS Working Papers on Corporate R&D and Innovation.

Collins, D. L., Román, F. J., and Chan, H. C. (2011). An empirical investigation of the relationship between profitability persistence and firms' choice of business model: evidence from the US airline industry. *Journal of Management Accounting Research*, 23(1), 37–70.

Connolly, R. A., and Hirschey, M. (2005). Firm size and the effect of R&D on Tobin's Q. *R&D Management*, 35(2), 217–223.

Cui, H., and Mak, Y. T. (2002). The relationship between managerial ownership and firm performance in high R&D firms. *Journal of Corporate Finance*, 8(4), 313–336.

Dasgupta, P., and Stiglitz, J. (1980). Industrial structure and the nature of innovative activity. *The Economic Journal*, 90(358), 266–293.

Ehie, I. C., and Olibe, K. (2010). The effect of R&D investment on firm value: an examination of US manufacturing and service industries. *International Journal of Production Economics*, 128(1), 127–135.

Erickson, G., and Jacobson, R. (1992). Gaining comparative advantage through discretionary expenditures: the returns to R&D and advertising. *Management science*, 38(9), 1264–1279.

Feinberg, S. E., and Majumdar, S. K. (2001). Technology spillovers from foreign direct investment in the Indian pharmaceutical industry. *Journal of International Business Studies*, 32, 421–437.

Freihat, Abdel Razaq Farah, and Raed Kanakriyah. "Impact of R&D expenditure on financial performance: Jordanian evidence." *European Journal of Business and Management* 9, no. 32 (2017): 73–83.

Gelei, A. (2003). Competitiveness: a match between value drivers and competencies in the Hungarian automotive supply chain. *Budapest University of Economic Sciences and Public Administration, Hungary*, 43–50.

Hana, H. (2004). The effete of advertising and R&D expenditures on the firm value. *An Empirical Analysis by Tobin's Q. Korean Accounting Review*, 2, 103–120.

Hausman, Jerry A. "Specification tests in econometrics." *Econometrica: Journal of the econometric society* (1978): 1251–1271.

Henard, D. H., and McFadyen, M. A. (2012). Resource dedication and new product performance: a resource-based view. *Journal of Product Innovation Management*, 29(2), 193–204.

Hsu, F.-J., Chen, M. Y., Chen, Y. C., and Wang, W. C. (2013). An empirical study on the relationship between R&D and financial performance. *Journal of Applied Finance and Banking*, 3(5), 107–119.

Insee, K., and Suttipun, M. (2023). R&D spending, competitive advantage, and firm performance in Thailand. *Cogent Business and Management*, 10(2), 2225831.

Kim, K., Hwang, J., Jung, S., and Kim, E. (2021). The optimal diversification strategy in pharmaceutical industry: balance-centred or hetero-centred? *Science, Technology and Society*, 26(2), 272–295.

Konak, Fatih, and Selçuk Kendirli. "Impact of R&D Expenses on Firm Performance: Emprical Evidence From the BIST Information Technology Index." *China-USA Business Review* 15, no. 5 (2016): 226–231.

Ministry of Chemicals & Fertilizer, Govt. of India, New Delhi (2008). Government of India Report on Indian Pharmaceutical Industry: World Quality Medicines at Reasonable Prices. Published by Department of Pharmaceuticals.

Monahan, S. J. (2005). Conservatism, growth and the role of accounting numbers in the fundamental analysis process. *Review of Accounting Studies*, 10, 227–260.

Mondal, P., and Sarkar, S. C. (2020). Impact of research and development investment on firms' profitability: a study on selected automobile companies in India. *International Journal of Multidisciplinary Studies*, 4(5), 9–18.

Morbey, G. K. (1988). R&D: its relationship to company performance. *Journal of Product Innovation Management: An International Publication of the Product Development and Management Association*, 5(3), 191–200.

Nam, J., Ottoo, R. E., and Thornton Jr. J. H. (2003). The effect of managerial incentives to bear risk on corporate capital structure and R&D investment. *Financial Review*, 38(1), 77–101.

Nath, P., Nachiappan, S., and Ramanathan, R. (2010). The impact of marketing capability, operations capability and diversification strategy on performance: a resource-based view. *Industrial Marketing Management*, 39(2), 317–329.

Paul, S. M., Mytelka, D. S., Dunwiddie, C. T., Persinger, C. C., Munos, B. H., Lindborg, S. R., et al. (2010). How to improve R&D productivity: the pharmaceutical industry's grand challenge. *Nature Reviews Drug Discovery*, 9(3), 203–214.

Poletti Hughes, Jannine. "R&D and dividend payments as determinants of corporate value in the UK: Empirical evidence after controlling for endogeneity." *International Journal of Managerial Finance* 4, no. 1 (2008): 76–91.

Ramirez, A. M., and Girdauskiene, L. (2013). Creation of knowledge and reverse logistics. empirical analysis from perspective of the resource based view theory. *Engineering Economics*, 24(5), 478–487.

Rao, J., Yu, Y., and Cao, Y. (2013). The effect that R&D has on company performance: comparative analysis based on listed companies of technique intensive industry in China and Japan. *International Journal of Education and Research*, 1(4), 1–8.

Selvarajah, K., and S. S. P. Sheena. "Innovation investment: an empirical study of technology public listed companies in Malaysia." *American Journal of Economics* 7, no. 1 (2017): 15–24.

Shen, K.-Y., Yan, M. R., and Tzeng, G. H. (2017). Exploring R&D influences on financial performance for business sustainability considering dual profitability objectives. *Sustainability*, 9(11), 1964.

Su, W., Song, X., and Guo, C. (2020). Research and development investment and operational performance of listed small and medium-sized enterprises in China. *Applied Economics*, 52(54), 5936–5948.

Vanderpal, G. (2015). Impact of R&D expenses and corporate financial performance. *VanderPal, GA (2015). Impact of R&D Expenses and Corporate Financial Performance. Journal of Accounting and Finance*, 15(7), 135–149.

Vurur, N. Serap, and Kenan Ilarslan. "Analysis of the relationship between R & D expenditure and profitability: a sample application from BIST." *Journal of Multidisciplinary Developments* 1, no. 1 (2016): 103–112.

Wang, Yanni, Rong Du, Kai S. Koong, and Weiguo Fan. "Effects of R&D policy choice on accounting performance and market value." *R&D Management* 47, no. 4 (2017): 545–556.

Xu, Jian, and Zhenji Jin (2016). Research on the impact of R&D investment on firm performance in China's internet of things industry. Journal of Advanced Management Science, 4(2). 112–116

9 Smart learning environments: an investigation of barriers to technology enabled blended learning in higher education

Surbhi Sethi[1,a], *Srishti Saxena*[1,b] *and Manju Singh*[2,c]

[1]Research Scholar, Department of Humanities and Social Sciences, Malaviya National Institute of Technology, Jaipur, India

[2]Professor, Department of Humanities and Social Sciences, Malaviya National Institute of Technology, Jaipur, India

Abstract

The COVID-19 pandemic has accelerated the shift towards online learning in higher education institutions (HEIs) in India, necessitating the adoption of technology-enabled blended learning (TEBL) to ensure uninterrupted educational delivery. This study employs a three-fold approach comprising a literature review, field observations, and the Fuzzy analytic hierarchical process (FAHP) to identify, validate, and prioritize the enablers of TEBL. The research involves 283 stakeholders from Indian HEIs, capturing their perceptions and priorities through a semi-structured questionnaire. The analysis highlights the priority weights of various barrier dimensions, including the students dimension, faculty dimension, institutions dimension, and infrastructure dimension. These dimensions encompass factors such as student engagement and adaptability, faculty preparedness and training, institutional policies and support, and the availability and quality of technological infrastructure. By addressing these barriers systematically, HEIs can more effectively implement TEBL strategies. The findings provide valuable insights for HEI stakeholders, helping to enhance the sustainability and effectiveness of educational practices through the use of TEL technologies. This research aims to foster a sustainable smart impact on higher education, ensuring that institutions can continue to deliver high-quality education in an increasingly digital world. The implications are significant, suggesting pathways for future policy and practice enhancements within HEIs.

Keywords: Barriers, blended learning approach COVID-19, fuzzy analytic hierarchy process, sustainability, technology-enabled learning

Introduction

The rapid transition to emergency remote teaching and learning has deepened the understanding of online education among stakeholders (Barnes, 2020). The pandemic, characterized by restrictions and social distancing measures, has accelerated the integration of technology into education, aligning with the long-standing concept of university 4.0 (Purcell, 2021). In India, the pandemic has accelerated digital change, resulting in active online course delivery, electronic evaluations, online interaction, and e-evaluation to enable continuous education (UNESCO, n.d.).While these measures have enhanced the understanding of online education, they have also underscored the limitations of traditional education systems dependent on physical presence. Online education has emerged as an effective alternative, prompting institutions worldwide to incorporate flexible teaching and learning methods (Teräs et al.,

[a]sethisurbhi.sethi3@gmail.com, [b]2020rhs9564@mnit.ac.in, [c]manjus.hum@mnit.ac.in

DOI: 10.1201/9781003564447-9

2020; Barnes, 2020). Today's students prefer a personalized learning experience that fosters personal development and lifelong learning capacity over traditional lecture formats. The pandemic has accelerated the shift from in-person to online education, prompting institutions to explore innovative teaching strategies as supplements to traditional methods (Bao, 2020; Shen et al., 2022). The shift from in-person to online education has brought forth challenges for instructors and students, encompassing issues such as digital infrastructure, administrative support, internet accessibility, availability of learning materials, platforms for delivering lessons, and the acquisition of tools and online licenses (Bao, 2020). In line with the *Aatma Nirbhar Bharat* program, the Indian government established the PM 'eVidya' program to support technology-driven education, emphasizing access and equity in the post-COVID era. Additionally, the MHRD introduced a virtual lab program allowing students to engage in laboratory sessions without the physical setup of instruments and equipment. Study Webs of Active Learning for Young Aspiring Minds (SWAYAM), the Indian chapter of massive open online courses (MOOC), was created to broaden access to high-quality education at affordable rates, aiming to bring transformative changes to overall educational outcomes (SWAYAM Government of India, All India Council for Technical Education, 2016). The university grants commission (UGC) has been urging colleges and universities to offer 20 percent of all courses online throughout a semester. To address language barriers in teaching and lecture delivery, these online platforms also provide course content in various Indian languages. technology-enhanced learning (TEL) emerges as an inclusive and sustainable solution to India's higher education demands. However, TEL is still in its early stages in India, facing challenges such as low levels of digital literacy, insufficient digital infrastructure, inadequate policies and procedures, and disparities in educational quality (Koneru, 2019).

Considering the reviewed literature, the study has the following research question and objective:

RQ1: What are the barriers to technology-enabled blended learning in higher education?

- To Identify and prioritize the barriers to technology-enabled blended learning in higher education

The purpose of this study is to close the gap identified by a meticulous literature review and employ extent analysis methods to investigate and prioritize the barriers to TEBL in university 4.0. The study involves a three-fold procedure for this purpose. Firstly, the barriers are identified from the literature and are classified into four dimensions – Students, faculties, institutions, and Infrastructure. Secondly, the barriers identified are validated in the field study. Thirdly, a fuzzy analytic hierarchical process is applied to prioritize the barriers for each dimension.

Literature Review

The COVID-19 crisis has underscored the vulnerabilities in the traditional educational system, heavily reliant on the synchronous physical presence of teachers and students. Instead of favoring direct, face-to-face lectures, contemporary learners prefer engaging in thought-provoking, experiential inquiries. The pandemic has compelled higher education institutions to incorporate more online and blended learning as part of emergency remote teaching, exacerbating existing disparities while expanding access to high-quality education (Swinnerton et al., 2020). Blended learning, in addition to its conventional subject selection approach, brings affordability, and flexibility in accessing course materials, and facilitates communication between students and instructors (Castle and McGuire, 2010). TEL not only enhances teaching-learning pedagogical methods but also improves learning outcomes by providing students with peer and instructor interaction beyond traditional classrooms (Kirkwood and Price, 2016). Amid the COVID-19 crisis, online education has emerged as a prominent practice, compelling institutions worldwide to incorporate diverse, flexible teaching and learning approaches (Huang et al., 2020). Challenges include digital infrastructure, administrative assistance, access to the internet, educational materials, delivering platforms, tools, and online licensing for instructors and students have arisen as a result of the move from in-person to online instruction (Hicks, 2021). The sudden shift to online instruction has left teachers and students with several problems, such as a lack of experience with online instruction, inadequate technology support, a shortage of educational resources, and a lack of a suitable learning environment (Bao, 2020). Given the circumstances, the higher education system must quickly restructure its digital policies, procedures, and infrastructure to meet the goals of the Sustainable Development Goals 4. This requirement motivates many institutions to investigate and use mixed learning strategies that are facilitated by technology (Elwyn et al., 2006). This strategy is in line with University 4.0's goal of improving educational outcomes and experiences by creating a self-paced, learner-cantered environment (Barnes, 2020). Government-imposed social restriction measures have increased the need for online educational and learning solutions (Purcell, 2021). Due to social alienation, the "work from home" paradigm has created difficulties for both educators and students, underscoring the critical necessity to engage and establish a connection with students in whatever way possible (Neuwirth et al., 2020). Both teachers and students faced difficulties with technological infrastructure, administrative assistance, connectivity to the internet, availability of educational materials, delivery channels, tools, and online licenses as they transitioned from in-person to online teaching and learning (Purcell, 2021; Barnes, 2020). Many universities have implemented virtual classrooms as a reaction to these difficulties, making use of Educational Technology (EdTech) tools such as online exams, video conferencing, and assessment instruments (Adekola et al., 2020). Flexible teaching-learning strategies, improved learner development, heightened faculty involvement throughout the course, and more student involvement with educational possibilities have all resulted from this shift (Neuwirth et al., 2020).

Students' Dimensions

Recognizing that e-learning aims to meet students' needs, the aspect of the student becomes crucial. TEL provides students with chances to interact with peers and instructors beyond traditional physical boundaries, thereby improving teaching methods and learning outcomes (Kirkwood and Price, 2016). Combining online technologies with traditional teaching methods is believed to enhance learning outcomes by maximizing the strengths and minimizing the weaknesses of each approach. This educational strategy can significantly enrich the teaching and learning experiences for diverse students outside the conventional, in-person instruction settings (Poon, 2013). To enhance learners' experiences, it is vital to identify and apply appropriate assessment techniques for both face-to-face and online environments (Laurillard, 2014). Research shows that students often lack experience in using technology for knowledge construction, as their familiarity usually involves using technology for entertainment and communication (Wang et al., 2014).

Faculty Dimensions

The role of faculty is essential in delivering lectures through a blended learning format (Garrison and Vaughan, 2008). HEIs can provide support and training to enhance the skills of faculty and staff, enabling them to innovate their teaching practices (Vaughan, 2007). Faculty training should shift from traditional didactic methods to more interactive, personalized learning environments that focus on tailored support. Research indicates that educators often lack experience in using technology for knowledge construction, as they are more accustomed to using it for entertainment and communication (Wang et al., 2014). Furthermore, the adoption of blended learning can make higher education institutions more flexible and adaptive, allowing them to respond swiftly to changes in a cost-effective way (Poon, 2013). Achieving this requires a comprehensive change in mindset and a redesign of the learning environment, including instructional methods, classroom technologies, and experiences, necessitating collaborative efforts from all faculty and staff members (Moskal et al., 2013).

Institutional Management Dimensions

The effective implementation and advancement of blended learning, a form of TEL, within HEIs, may necessitate a comprehensive vision firmly rooted in the institutions' policies, processes, and practices related to teaching and learning in blended environments (Bates and Sangrà, 2011). This approach exhibits significant potential for increased engagement with scalability among learners, academia, and institutions (De Freitas et al., 2015). It holds the capacity to reshape educational policies, directing pathways toward a more inclusive knowledge world (Vuorikari and Castaño Muñoz, 2018; Rasheed, 2020). Therefore, it is imperative to develop focused approaches at the institutional level to enhance student experiences with the active involvement of all stakeholders. Advancements in technology, such as big data analytics, deep learning, and the Internet of Things (IoT), are influencing teaching and learning across all levels of the education system (Castle et al., 2019).

Infrastructure Dimensions

The effective execution and enhancement of blended learning, a form of TEL, within HEIs, might require a well-defined vision that is thoroughly integrated into the institutions' policies, processes, and practices for teaching and learning in blended environments (Bates and Sangrà, 2011; Bordoloi, 2021). This method holds significant promise for increasing engagement and scalability among students, faculty, and institutions (De Freitas et al., 2015; Garrison, 2018). It can redefine educational policies, paving the way for a more inclusive knowledge society (Vuorikari and Castaño Muñoz, 2018). Therefore, developing focused strategies at the institutional level to improve student experiences with the active participation of all stakeholders is essential. Technological advances, such as big data analytics, deep learning, and the IoT, are influencing teaching and learning at every level of the education system (Castle et al., 2019).

Methodology

The study involves a three-fold procedure for this purpose. Firstly, the barriers are identified from the literature and are classified into four dimensions – students, faculties, institutions, and infrastructure. Secondly, the barriers identified are validated in the field study. The survey involved the implementation of a validated & reliable data collection tool. The tool is validated in a pilot study using the Delphi technique, and the perceptions of the stakeholders toward the barriers are captured. Finally, a FAHP is applied to prioritize the barriers for each dimension. The problem of strategizing a robust TEL ecosystem to reduce the negative impacts of these barriers requires ranking the barriers as per their severity. Therefore, the problem can be classified as a multiple-criteria decision-making (MCDM) problem as there is the existence and co-existence of various criteria and alternatives. The Analytic Hierarchical Process is a well-known MCDM procedure and has been used popularly in scenarios where multiple independent factors affect the decision-making process of stakeholders (Saaty, 2008). The calculations of the pair-wise comparison weight matrix of the barrier dimensions and their contributing factors are based on the expert's priorities and perceptions captured in a field study. The study involves interviewing 283 stakeholders. The minimum sample size is calculated at 283 to have a 95% confidence level and a margin of error of 0.05. A semi-structured questionnaire was administered to collect the responses from the stakeholders of the HEIs – students, faculty, management and administrative staff, and industry. The relative importance is determined on a 9-point "Saaty's Fundamental Scale" (Saaty, 2008). The Questionnaire tool's validity and reliability were evaluated using the Delphi technique. The research employs focused group discussions for validating the barriers identified from the literature with field observations. The numerical judgments at different hierarchy levels are used to construct the pairwise matrices for the attributes (Chang, 1996). These matrices are normalized eigenvectors that yield a natural measure of consistency. To achieve the robust prioritization of the barriers, the FAHP has justified utility (Bozbura et al., 2007). The first and foremost step is to develop a Hierarchical structure as shown in Figure 9.1. Then the relative importance is determined, and the pairwise comparison matrix is utilized. A set of fuzzy numbers (a, b, c) in triangular form is employed to

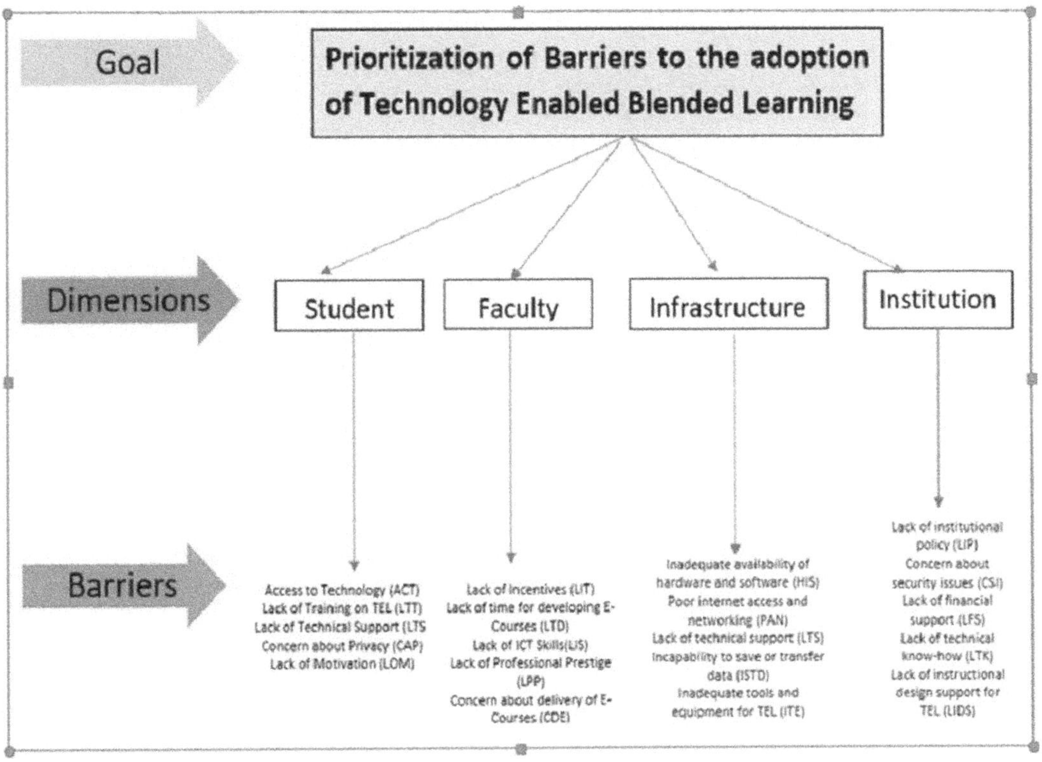

Figure 9.1 Hierarchy of barrier dimension and contributing factors
Source: Authors Compailation

facilitate decision-making (Husain et al., 2021).

Analysis and Results

Step I: Construct the pairwise comparison matrix for respective barrier dimensions and their factors in terms of the TFN.

A normalized pairwise comparison matrix and a fuzzy number matrix establish the relative significance of different attributes or criteria concerning the goal. The weights for the fuzzified pairwise comparison matrix can be derived using the geometric mean method (Buckley, 1985).

Step II: Calculating fuzzy synthetic extent

Fuzzy synthetic extent shown in Table 9.1,

$$S_i = \sum_{i=1}^{m} M_g^j \varphi \left[\sum_{=1}^{n} \sum_{j=1}^{m} M_{gi}^j \right]^{-1} i$$

Step III: Determining the degree of possibility of supremacy for two TFNs. The degree of possibility is calculated to compare the fuzzy numbers with each other

Let there be two TFNs $M_1 = (l_1, m_1, u_1) \geq M_2 = (l_2, m_2, u_2)$

$$V(M_2 \geq M_1) = \mathrm{Sup}\,[\min(\mu_{M_1}(x), \mu_{M_2}(y))]$$

And is equivalent to

$$V(M_2 \geq M_1) = hgt(_1 \cap M_2) = \mu_{M_2}(d)$$

$$\mu_{M_2}(d) = \begin{cases} 1, \text{ If } m_2 \geq m_1 \\ 0, \; l_1 \geq u_2 \\ \dfrac{l_1 - u_2}{(m_2 - u_2) - (m_1 - l_1)}, \; otherwise \end{cases}$$

The degree of possibility is shown in Table 9.2. The degree of possibility is a major component in calculating the weights of each barrier.

Step IV: Comparing the degree of possibility of a convex fuzzy number is more significant than k convex

Table 9.1 Fuzzy synthetic index for barrier dimension.

Barrier dimension number	Barrier dimension	Fuzzy synthetic index
I	Student's dimension (SD)	(0.741, 0.989, 1.273)
II	Faculty's dimension (FD)	(0.713, 0.984, 1.304)
III	Institute's dimension (SI)	(1.26, 0.977, 1.343)
IV	Infrastructure dimension (SID)	(0.982, 0.99, 0.858)

Source: Authors' computation

Table 9.2 Degree of possibilities.

Degree of possibility	V(SD)	V(FD)	V(SI)	V(SID)
V(SD)	-	0.9879	0.9845	1.0086
V(FD)	0.9911	-	1	1
V(SI)	0.9804	0.9890	-	1
V(SID)	1.0086	1.0431	1	-

Source: Authors' computation

fuzzy numbers. The minimum degree of possibility is calculated for each barrier dimension

d'(SD) = min (0.9879, 0.9845, 1.0086) = 0.9845

d'(FD) = min (0.9911, 1, 1) = 0.9911

d'(SI) = min (0.9804, 0.9890, 1) = 0.9804

d'(SID) = min (1.0086, 1.0431, 1) = 1.0086

Hence W_0 = (0.9845, 0.9911, 0.9804, 1.0086)

Step V: Normalizing the weight vectors

The normalized weight vector is of the student's dimension (SD), faculty's dimension (FD), institution's dimension (SI), and infrastructure dimension (SID) is W_G = (0.3303, 0.3325, 0.3289, 0.3384).

Step VI: Overall score calculations for each barrier dimension and its factors

The pairwise and fuzzified comparison matrix yields the priority weights of each barrier dimension and its factors. The weights of the barrier dimension are listed in the table under the column level I priority, and the weights corresponding to the factor dimension are mentioned at level II priority. The final priorities used for ranking the severity of the barriers are obtained as the product of the level I and level II priorities shown in Table 9.3.

Table 9.3 Severity ranks of the barrier dimensions and the contributing factors.

Barrier dimension	Level I priority	Barrier factors	Level-II factors	Final priorities	Prioritization through FHP
Student's dimension (SD)	0.9845	ACT	0.48	0.47	3
		LTT	0.31	0.31	5
		LTS	0.05	0.05	20
		CAP	0.11	0.11	14
		LOM	0.05	0.05	19
Faculty's dimension (FD)	0.9911	LIT	0.47	0.47	4
		LTD	0.23	0.23	6
		LIS	0.06	0.06	17
		LPP	0.13	0.13	11
		CDE	0.11	0.11	13
Institution's dimension (SI)	0.9804	LIP	0.58	0.57	1
		CSI	0.14	0.14	10
		LFS	0.13	0.13	12
		LTK	0.09	0.09	15
		LIDS	0.06	0.06	18
Infrastructure's dimension (SID)	1.0086	HIS	0.48	0.48	2
		PAN	0.16	0.16	7
		LTS	0.15	0.15	9
		ISTD	0.15	0.15	8
		ITE	0.06	0.06	16

Source: Authors' computation

The fuzzy analytics hierarchy process yields the priority weights of barriers dimensions: students dimension (SD), faculty dimension (FD), institutions dimension (SI), and infrastructure dimension (SID). The priority weights thus obtained are 0.9845, 0.9911, 0.9804, and 1.0086. The severity of the barriers can be judged by using priority weights and can be ranked as SID>FD>SD >SI, where the ">" indicates the order severity of barrier dimensions.

The results obtained yield the following insights:

1) The infrastructure dimension has maximum weightage, whereas the institution dimension has a minimum weightage.
2) Considering the level two priorities top four barriers in terms of severity are lack of institutional policy (LIP), inadequate availability of hardware and software (HIS), access to technology (ACT), and lack of incentives (LIT).
3) Ranking of sub-criteria of infrastructure dimension is inadequate availability of hardware and software (HIS)> poor internet access and networking (PAN) > incapability to save or transfer data (ISTD) > lack of technical support (LTS) > Inadequate tools and equipment for TEL (ITE).
4) Ranking of sub-criteria of faculty dimension is lack of incentive (LIT) > lack of time for developing e-courses (LTD) > lack of professional prestige (LPP) > concern about delivery of e-courses (CDE) > lack of ICT skills (LIS).
5) Ranking of sub-criteria of student dimension is Access to technology (ACT) >lack of training on TEL (LTT)> concern about privacy (CAP) > lack of motivation (LOM) > lack of technical support (LTS).
6) Ranking of sub-criteria of institution dimension is lack of institution policy (LIP)> concern about security issues (CSI) > lack of financial support (LFS) > lack of technical know-how (LTK) >lack of instructional design support for TEL (LIDS).
7) According to the FAHP analysis, the overall ranking of the factors is LIP> HIS> ACT> LIT> LTT> LTD> PAN> ISTD> LTS> CSI> LTT> LFS> CDE> CAP> LTK> ITE> LIS> LIDS> LOM> LTS. This prioritization underscores the critical importance of establishing clear institutional policies to support online learning initiatives, ensuring adequate infrastructure and technology access, and providing incentives to motivate stakeholders.

Strategizing the mitigation procedure for overcoming these barriers is essential.

Policy Implications

This study has several policy implications for researchers, academicians, industry practitioners, community organizations, and Edtech leaders. First, educational institutions should prioritize the development and implementation of comprehensive policies that support and regulate online learning initiatives. These policies should address issues such as course delivery methods, technological requirements, academic integrity, and student support services. Next, acknowledging the significance of infrastructure (HIS) and technology accessibility (ACT), decision-makers ought to designate resources to guarantee the provision of essential hardware, software, and internet access for both educators and learners. This may involve investing in upgrading existing infrastructure and providing subsidies or grants to facilitate access to technology for economically disadvantaged students. Third, To address the barrier of lack of incentives (LIT), institutions should consider implementing incentive programs to motivate faculty members to engage in e-learning development and delivery. This could include recognition for innovative teaching practices, financial rewards, or opportunities for professional development and career advancement. Fourth, Given the importance of technical skills (LIS) and training on technology-enhanced learning (LTT), institutions should invest in faculty development programs to enhance ICT skills and pedagogical knowledge related to online teaching and learning. Additionally, providing ongoing technical support (LTS) and instructional design support (LIDS) can help faculty members and students navigate challenges and maximize the effectiveness of online learning experiences. Fifth, the policymakers and think tanks have begun to rewire for new pedagogies, technology, and methods of teaching and learning. The pandemic experience has provided a creative chance to change the educational system by putting more of an emphasis on adaptive learning platforms, which will help combat stigmatizing individuals who are excluded and advance inclusive sustainable, and smart learning environments for HEIs. By addressing these policy implications, educational institutions can better overcome the identified barriers and create a supportive environment conducive to the successful implementation and adoption of online learning initiatives.

Conclusion, Limitations, and Future Research Agenda

The study investigated and prioritized the barriers to TEBL in higher education. The study involved a three-fold procedure for this purpose. Firstly, the barriers were identified from the literature and were classified into four dimensions – Students, faculties, management & administration, and infrastructure. Secondly, the barriers identified were validated in the field study. A FAHP is applied to prioritize the barriers for each dimension. The barriers to TEL were identified based on a limited literature survey and were further supported by capturing the perceptions of the stakeholders by administering a sample survey. The Study lacks the identification of the enablers and their interaction effects with the implications proposed. Future studies can employ the Technique for Order Preference by Similarity to the Ideal Solution (TOPSIS) procedure to map the systematic effective strategies with the enablers that can further be adopted to enhance the skills of all stakeholders for a TEBL. The research will address educational disruption and build up a long-term resilient education system with TEBL, a pathway for quality higher education and knowledge democracy. Educational institutions have realized that the system is much more than just content, assessments, and evaluation but also social justice, access equality, quality, emotional relationships, and lifelong learning, as emphasized by SDG4. Indeed, this pandemic has become a growth opportunity for sensitizing the concept of the education system through more sustainable and inclusive integration of TEBL. This education model opens the way for critical investigation of the existing practices and how they can be further embraced to improve access, creation, and refinement of knowledge in post-covid India.

References

Adekola, J., Dale, H., and Gardiner, M. (2020). A conceptual model for technology-enhanced learning at De Montfort University. *Journal of Educational Technology Systems*, 49(2), 204–222.

Bao, W. (2020). COVID-19 and online teaching in higher education: a case study of Peking University. *Human Behavior and Emerging Technologies*, 2(2), 113–115.

Barnes, C. (2020). The effects of emergency remote teaching on students and educators. *Journal of Online Learning*, 14(1), 23–34.

Bates, A. W. T., and Sangrà, A. (2011). Managing Technology in Higher Education: Strategies for Transforming Teaching and Learning. San Francisco: Jossey-Bass.

Bordoloi, R. (2021). Challenges in implementing e-learning in higher education institutions. *Journal of Education and Research*, 9(1), 45–58.

Bozbura, F. T., Kahraman, H., and Erdoğan, G. (2007). Fuzzy multi-criteria decision making: a review of the literature and some issues. *International Journal of Approximate Reasoning*, 18(4), 423–438.

Buckley, J. J. (1985). Fuzzy hierarchical analysis. *Fuzzy Sets and Systems*, 17(3), 233–247.

Castle, S. R., and McGuire, K. B. (2010). An analysis of student self-assessment of online, blended, and face-to-face learning environments: implications for sustainable education delivery. *International Education Studies*, 3(3), 36–47.

Chang, D. Y. (1996). Applications of the extent analysis method on fuzzy AHP. *European Journal of Operational Research*, 95(3), 649–655.

Castle, S. R., Myers, J. W., & Waldman, L. R. (2019). The Impact of Technology on Education: A Review of Current Trends and Future Directions. Journal of Educational Technology Systems, 48(2), 123–145. https://doi.org/10.1177/0047239519876593

De Freitas, S., Morgan, M., and Gibson, H. (2015). Will MOOCs transform learning and teaching in higher education? engagement and course retention in online learning provision. *Online Learning Journal*, 19(2), 15–25.

Elwyn, G., Edwards, A., and Grol, R. (2006). Shared decision making: a model for clinical practice. *Journal of General Internal Medicine*, 21(1), 41–46.

Garrison, D. R., & Vaughan, N. D. (2008). Blended learning in higher education: Framework, principles, and guidelines. John Wiley & Sons.

Garrison, D. R. (2018). Understanding blended learning in higher education: opportunities and challenges. *The Internet and Higher Education*, 18(1), 29–37.

Hicks, M. (2021). Overcoming the challenges of online teaching: strategies and tools for success. *Educational Technology Research and Development*, 69(3), 365–378.

Huang, R., Liu, D., and Xia, S. (2020). Online education in the post-COVID era: new opportunities and challenges. *Educational Research and Reviews*, 15(6), 299–307.

Husain, H. A., Yusof, M. F. M., Selamat, M. H., Abas, N., and Taib, M. (2021). A fuzzy analytic hierarchy process (FAHP) approach for identifying the critical success factors in e-learning implementation in higher education. *Journal of Information and Communication Technology*, 20(2), 177–202.

Kirkwood, A., and Price, L. (2016). Technology-enhanced learning and teaching in higher education: what is 'enhanced' and how do we know? a critical literature review. *Learning, Media and Technology*, 39(1), 6–36.

Koneru, I. (2019). Technology-enhanced learning: policies and practices in India. *Journal of Educational Technology*, 15(2), 45–57.

Laurillard, D. (2014). Teaching as a Design Science: Building Pedagogical Patterns for Learning and Technology. New York: Routledge.

Moskal, P. D., Dziuban, C., and Hartman, J. (2013). Blended learning: a dangerous idea? *The Internet and Higher Education*, 18, 15–23.

Neuwirth, L. S., Jović, K. J., and Mukherji, S. M. (2020). Reimagining higher education during and post-COVID-19: challenges and opportunities. *Journal of Adult and Continuing Education*, 26(2), 141–156.

Poon, J. (2013). Blended learning: an institutional approach for enhancing students' learning experiences. *Journal of Online Learning and Teaching*, 9(2), 271–288.

Purcell, W. M. (2021). University 4.0: a blueprint for disruption in higher education. *Technology and Education Journal*, 36(4), 45–56.

Rasheed, A. (2020). Challenges of e-learning during the COVID-19 pandemic in Iraq. *Journal of University of Babylon for Humanities*, 28(6), 143–156.

Saaty, T. L. (2008). Decision Making with the Analytic Hierarchy Process. Pittsburgh: RWS Publications.

Shen, J., Wang, D., and Zhang, Y. (2022). Blending traditional and online teaching: a new model for post-pandemic education. *Educational Innovations*, 18(3), 78–89.

Government of India, All India Council for Technical Education (2016). SWAYAM:

Study Webs of Active Learning for Young Aspiring Minds. Retrieved from https://swayam.gov.in.

Swinnerton, B., Purcell, W. M., and Barnes, C. (2020). Emergency remote teaching in higher education: new directions in teaching and learning. *Technology and Education Journal*, 36(4), 23–34.

Teräs, M., Suoranta, H., Teräs, J., and Curcher, M. (2020). Post-COVID-19 education and education technology 'solutionism': a seller's market. *Postdigital Science and Education*, 2(3), 863–878.

UNESCO (n.d.). Education: From Disruption to Recovery. Retrieved from https://en.unesco.org/covid19/educationresponse

Vaughan, N. D. (2007). Perspectives on blended learning in higher education. *International Journal on E-Learning*, 6(1), 81–94.

Vuorikari, R., and Castaño Muñoz, J. (2018). Research evidence on the use of learning analytics: implications for education policy. *European Commission*. https://doi.org/10.2791/955210

Wang, S.-K., Huang, H. H., Hsu, S. J., Campbell, L. O., Coster, D. C., and Longhurst, J. (2014). Using digital technologies to enhance classroom interaction. *Journal of Educational Technology and Society*, 17(2), 162–174.

10 Enabling sustainable progress in rural regions through an intelligent management system for enhanced business profitability

Prasanta Kumar Parida[1,a], *Monalisha Chakraborty*[2,b] *and Jyotirmayee Pati*[2,c]

[1]Associate Professor at School of Rural Management, KIIT University, Bhubaneswar, Odisha, India

[2]PhD Scholar at School of Rural Management, KIIT University, Odisha, India

Abstract

Throughout the long history of agriculture, organic farming has given way to inorganic farming at every turn. The advancement of technology has made it possible to modernize agriculture. To address the drawbacks of conventional methods of detecting targets in images, such as harvesting insects, a method utilizing deep learning (DL) that emphasizes identification and detection is suggested. It also looks into agricultural data oversight and management as well as the productivity of DL technology. The design of this system emphasizes the main clause's strong technological foundation and reorganizes the latter portion of the phrase for clarity. For rural businesses, especially those with limited resources or a smaller workforce, it could be challenging to cover the expenses of putting in place and keeping up an intelligent management system. To solve these issues, we suggested using an adaptive spider monkey swarm optimized enhanced deep neural network (ASMSO-EDNN), which was created to handle several issues about sustainability and rural development. The data is compiled using image target detection with the MPD dataset. Normalization techniques are part of data preparation, and they are essential to developing policies and making well-informed decisions in rural areas. The principal components are extracted using principal component analysis (PCA). An examination of the precision-recall curve indicates that the proposed technique reduces the rate of false positives and ignored positives. The recommended approach to pest control using blockchain takes pattern recognition into account, which is advantageous for both the long-term viability of rural communities and the understanding of farming data processing.

Keywords: Adaptive spider monkey swarm optimized enhanced deep neural network, business profitability bank, development of sustainability, principal component analysis, regional governance

Introduction

The need to promote sustainable development (SD) in rural areas has never been urgent in a time of rising environmental consciousness and technological advances. To increase corporate profitability while conserving these communities' ecological and social fabric, it is necessary to design an intelligent management system that can harness data-driven insights and creative solutions. This initiative will support rural enterprises, promote economic expansion, and maintain environmental harmony.

Financial System

A functioning and expanding financial system is necessary for an economy. Typically, it is believed that the financial system's

[a]prasanta.parida@ksrm.ac.in, [b]monalishachakr11@gmail.com, [c]jyotirmayee.pati17@gmail.com

DOI: 10.1201/9781003564447-10

(FS) banking sector has the most sway. The banking industry has faced significant obstacles since the country's independence in 1947 and several political changes. A lack of personnel and financial resources, governmental interference, and political instability have hampered industry expansion. Despite these difficulties, the banking industry has expanded over the past 20 years, partly due to the economy's 5.2% annual growth. A functioning and developing FS is necessary for an economy. The financial sector is considered to have the most influence inside an FS (Harris et al., 2019). Intellectual capital (IC) is viewed as a significant contribution to businesses in the information economy rather than tangible assets in terms of improving company competitiveness and creating value. According to resource-based theory, each organization's resources are exclusive. IC a recognized strategic resource is connected to securing 161a competitive edge and better performance via value creation. As a result, businesses must comprehend, discover, create and use IC, which may provide them with a competitive edge (Xu and Liu, 2020). Figure 10.1 depicts the schematic representation of business profitability.

Business Intelligence (BI)
Businesses may use BI to acquire crucial data from many disorganized informational sources and turn it into actionable knowledge. This will help them make better policy decisions and increase organizational effectiveness and profitability. Every business has obstacles while using BI and making decisions, including planning shortcomings, resource underutilization, ineffective resource usage, and risk-taking propensities (Niu et al., 2021). BI describes a company's capacity to use the data obtained through regular activities. The BI may improve organizational effectiveness by seeing fresh possibilities, indicating potential dangers, offering extra company knowledge and strengthening decision-making frameworks. BI uses current and historical data to conduct company operations, fuel customer demand and increase productivity. Business analytics examines previous data. For BI systems, internal and centralized business data are the primary information sources (Turban, 2022).

Rural Regions
The United States Environmental Studies Program's 2018 National Climate

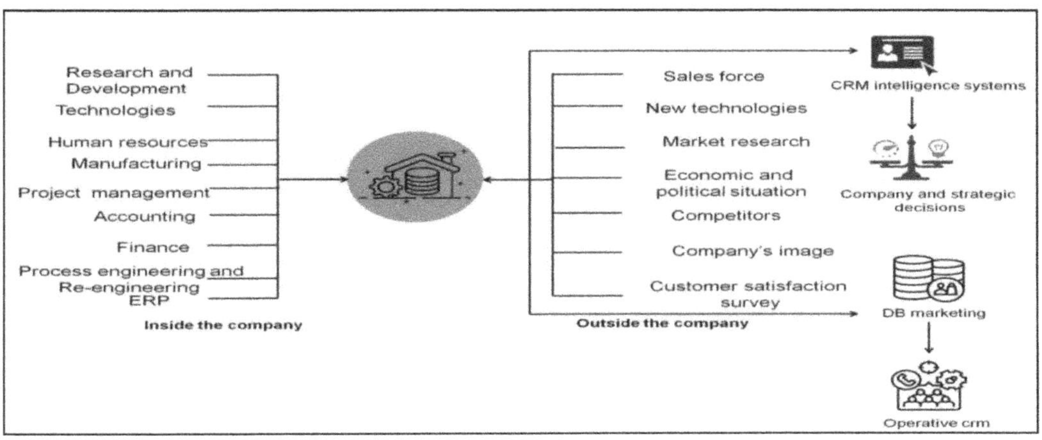

Figure 10.1 Business profitability
Source: VAIC model of Nazari and Hermans(20017.modified for my research for this research)

Assessment notes, additional challenges to rural areas' economic vitality, agricultural and livestock health, and productivity due to global warming (Charania and Li, 2020). Over the following decades, some regions might experience favorable conditions for increased or alternative crop productivity. Total yields from essential United States products are predicted to decrease owing to rising temperatures, disease outbreaks, and potential changes related to water availability, soil erosion, and insects.

Balancing Sustainability

According to the World Commission on Environment and Development, SD is the process of meeting current requirements without sacrificing the capacity of the demands of successive generations will be addressed. Numerous studies have been done regarding the idea of SD investigation at both the macroeconomic and microscopic levels in various academic studies (White, 2023). SD is quite popular since it sets common patterns for the company and non-business activities and all current research and development, both academic and practical. To develop and implement an intelligent management system to boost company revenue while promoting long-term growth in rural areas.

Contributions to the Study

- A system increases firm profitability through improved resource allocation, decreased waste, and increased production. Rural regions experience an increase in economic growth.
- Rural areas may enhance technological literacy and creativity by introducing cutting-edge technology and data-driven decision-making.
- The system can promote enhanced governance and policy frameworks adapted to the particular requirements of rural regions by offering more significant insights into rural development and agricultural practices.

There is an established research summary: Section II provides a list of related works, Section III explains the methodology, Section IV discusses the findings and Section V draws a conclusion.

Literature Reviews and Related Work

The study Fina et al. (2019) examined the economics of photovoltaic (PV) systems and the optimal implementation capability for energy communities (ECs). Four distinct settlement patterns, from urban to suburban to historical to rural, are outlined to provide a wide range of results. Many nations are updating their legislative frameworks to permit sharing without regard to building limitations. In addition, examining the research areas that come under the Industry 4.0 framework and taking into account the points above Khan et al. (2021) presented a map of the vast field of SD. Industry 4.0 is credited as playing a significant role in the current digitization era. Its implications for SD have been highlighted by the triple bottom line, sustainable business models, and the economy of circularity. The study Tien et al. (2020) covered the ideas, challenges, sustainability, and tactical elements of social companies and social entrepreneurship.

This study showed that the two current forms of entrepreneurship when compared to their nature, inspiration, goal, and orientation toward SD, might ally to bridge social and economic capital for business environmental and social SD. However, the essay demonstrated that, in contrast to economic business ownership, social enterprise lacks and requires a calculated strategy and understanding, as well as its complete reflection in both theoretical discussions and corporate realities. This is due to the sophistication of social entrepreneurship compared to commercial entrepreneurship. To assess the empirical effects of a strategic management system on SD, the balanced scorecard (BSC) was employed as a theoretical framework and the success of the company as an intervening component in the Study's Rafiq et al. (2020) effort. To address the investigation's objectivity, the study used a positivist research paradigm. A quantitative strategy was the most effective method for reaching the study's research goals since it is an empirical and objective investigation. Surveys were employed to collect data analysis. According to the research, there have been initiatives to reduce the gap between theories about how businesses may enhance urban resilience and the actual measures they employ and reveal (Acuti et al., 2020). Although enterprises play a significant role in strengthening urban resilience, there are uncertainties regarding the possible tactics they may use. Given that the idea of resilience is acknowledged in various SD Goals (SDGs) both expressly and implicitly.

The study Nayal et al. (2022) used Orchestration of Resources, a detailed rendition of the Expanded Utilizing Resources, as well as Dynamic Capability Theory to define the relationship between flexibility and supply chain (SC) business performance under th.e circular economy (CE) context. A questionnaire-based survey was used to gather data from Indian manufacturing companies, and structural equation modelling (SEM) was used to analyze the results. Practitioners and scholars have focused on the digitization of the SC in recent years to assist organizations in attaining a circular SC and address issues with the business environment. The Study Han et al. (2020) discussed the product as a service model and focused on service-led growth as a cutting-edge company strategy in the CE.

In reaction to technological shifts, the economy and the environment, a business model must be adaptable to change with the market. The appropriate business model for resource use is becoming crucial for the success of businesses in resource-constrained environments. Using a fresh approach to business modelling in a CE is vital. The Study Nayal et al. (2021) looked at how the adoption of blockchain technology (BCT) affects the effectiveness of sustainable supply chains (SSCP). This research examined the adoption of BCT in connection to various factors, including green and lean practices, regulatory support, supply chain integration, cost and support from upper management, innovation capabilities, and internal and external environmental circumstances. A sample of 316 respondents from Indian agriculture supply chain (ASC) enterprises was subjected to SEM.

The ASC must resolve its issues to attain SD. The article Ciliberto et al. (2021) explored initiatives to integrate lean manufacturing with ethical production, stressing the chance to participate in reverse logistics and the Industry 4.0 system as a stepping stone to realizing CE objectives. The Study aims to investigate the relationships between lean production, industry 4.0, and sustainable manufacturing to demonstrate, why organizations should embrace these practices from an SD standpoint. The Study Gupta et al. (2022) examined the links between Industry 4.0, business methods and operational performance. An investigative model is developed and assessed using the SEM approach. The study demonstrated that SMEs can improve sustainability in manufacturing operations by implementing Industry 4.0 practices, using advanced robotics that decreases human involvement, virtual reality, intelligent logistics that adapt to changes in manufacturing capability and dynamic modelling methods. In the current competitive market, to understand the structure of Industry 4.0 and adopt ethical procedures for improved procedures, SMEs need to employ advanced equipment. Table 10.1.

Approach and Methodology

The list of tasks that follow is completed using the indicated approach. We first collected the data set and used Z score

Table 10.1 A Summary of the literature with results and limitations.

Ref No	Objective	Result	Limitation
(Fina et al., 2019)	In contrast with isolated constructions, the study examined the economics of photovoltaic (PV) systems and the optimal implementation capability for energy communities (ECs).	According to the findings, creating ECs enhances the profitability of installing properly sized solar power plants as opposed to the case when constructions are considered individually. The potential for cost savings increases with the diversity of the load profiles through increasing interaction impacts.	Complex laws, constrained scalability and uneven advantages among participants in diverse settlement patterns may make PV sharing in energy communities less profitable.
(Khan et al., 2021)	The period of digitalization has been attributed in large part to Industry 4.0. Due to its significance for SD from the viewpoints of the triple bottom line, ethical business practices, and CE, it has received a lot of attention.	Industry 4.0 acceptance execution and ethical supply networks are the focus of two new research themes: the CE and ethical business practices.	The comprehensive integration of Industry 4.0, SD, and triple-bottom-line viewpoints has limitations due to data heterogeneity, conceptual variety, and changing standards of sustainability.
(Rafiq et al., 2020)	The contest encourages teams to develop a long-term, distinctive and easily adaptable strategic management system to boost team productivity. In the past, organizations employed a company's performance system of management to accomplish their financial objectives, but they learned that non-financial indicators were equally significant.	The effectiveness of the firm is used to be an influencing factor within the present research to assess the effect of SMSs on SD. Smart-PLS has been used in the Study to explore and analyze outcomes.	The key motivation for firms to improve their strategic management system is to increase profitability and achieve long-term success. The business world is forced by the fierce competition in today's dynamic environment to pursue efficient strategic systems that can enable firms to function.
(Han et al., 2020)	Information and communication technology (ICT) formed the foundation for service-led growth, a sector of development that has grown in importance throughout the fourth industrial century. This study emphasized service-led growth as a cutting-edge commercial approach for the CE and provided the product as a model for service.	Due to the development of new intelligent connected products such as the fifth-generation network technologies Internet of Things, this business model can be successful. The CE for businesses, customers and the environment, as well as cost and convenience, are critical components of this new business model.	Implementing circular economy business models with a servitization focus has been constrained by market acceptability, resource limitations, and regulatory issues.
(Gupta et al., 2022)	In a competitive environment, SMEs leverage modern technology to comprehend new industry	This outcome will assist businesses in achieving operational excellence with increased	Data accessibility, emerging technology, and organizational opposition are obstacles

Ref No	Objective	Result	Limitation
	trends and continual actions for the best results. It looks at the relationships connecting Industry 4.0, moral business conduct and business efficiency.	productivity and efficiency. As a result, SMEs must concentrate on sustainable production techniques.	to creating a structural paradigm for sustainable operations provided by Industry 4.0.
(OECD, 2018)	In rural communities to benefit fully from the most recent technological and digital developments, they require robust communications networks. Using Internet-based digital services via a range of technologies—such as satellite, TV white spaces, high-frequency 4G LTE, Digital Subscriber Line (DSL), or TV white spaces—is made possible by broadband access.	Broadband access is essential for social and economic interaction and has a favourable impact on local labour market results, firm productivity, and the number of businesses in an area.	Within OECD nations, disparities in internet connectivity continue to exist by region. The term "urban-rural digital divide" refers to the relative lack of digital connectivity that rural settlements experience in comparison to urban areas.
(Global Workplace Analytics, 2018); (Clark, 2018)	The shifts in working practices will be advantageous for rural areas. Compared to their metropolitan counterparts, they provide more natural amenities and a cheaper cost of living (Clark, 2018). Before the COVID-19 pandemic, medium-sized and smaller towns in the US had a higher percentage of teleworkers than larger cities (Global Workplace Analytics, 2018).	Some OECD nations have already looked into teleworking as a policy tactic to strengthen rural economies before the current crisis. For instance, teleworking has been employed by Japan as a public policy to boost the employment rate of the disabled population and support local revitalization. Similar to this, the US has supported projects to construct outposts and creative areas where individuals can operate from a distance.	The 2020 lockdown measures showed that, as a result of the new population's migration to rural areas and changes in economic composition, capital regions had the greatest rates of distant working.

Source: Based on Primary data from field study

normalization to pre-process the data. The pre-processed data was used to extract features using independent component analysis. Employing the use of PCA, choose the best characteristics. Figure 10.2 depicts the stages, and the following parts give further information on each step in greater detail.

Principal component analysis, or PCA, is a dimensionality reduction approach that is commonly used to reduce the dimensionality of large data sets. It does this by taking a large collection of variables and breaking them down into a smaller one that still has much of the information in the bigger set. SEM stands for scanning electron microscope. An example of a microscope that uses electrons rather than light to produce images is the SEM. Since its development

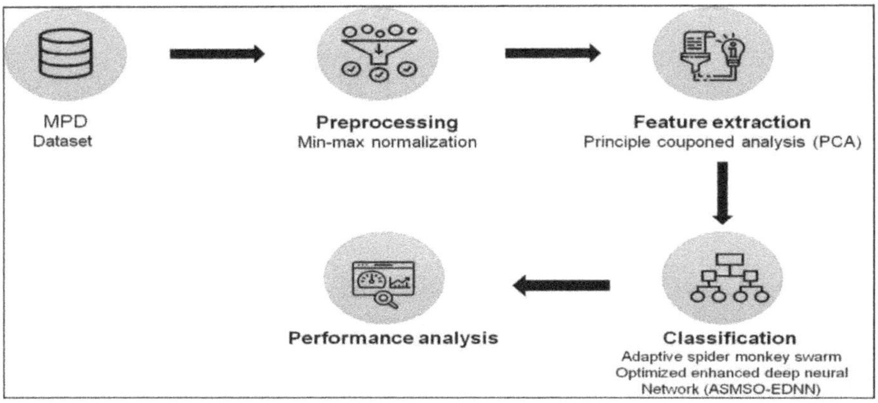

Figure 10.2 Methodology
Source: Schematic diagram prepared for this research

in the early 1950s, scanning electron microscopes have provided new opportunities for study in the domains of physical science and medicine.

Data Set

The 10.2 different pest species that make up IP102's taxonomy of insects. Due to their insufficient ability to harm certain crop categories, many of these pest insects do not need to be avoided in practical farming. Furthermore, several articles have observed that the two primary crops that are affected by pests are wheat and rice (Wang et al., 2021). Consequently, update the IP102 insect classification and concentrate on problems affecting 4 distinct crop types. In addition, 14 pest groups are identified, which are divided into four super-classes that correspond to the four most cultivated crops for cultivation. The Rice Plant Hopper (RPH), a subclass of one of the super-classes, harms rice crops.

Date Pre-Processing Using Z-Score Normalization

To set a dataset's mean and standard deviation to 0 and 1, respectively, statistical "standardization," also called "Z-score normalization," is applied. Z-score describes the steps taken to get a dataset ready for use. Analysis is difficult because data arising from varying input quality is gathered in an unprocessed state. As a result, specific techniques are employed to convert data into a trustworthy dataset. This approach—dubbed "data pre-processing"—uses the normalized data pre-processing model without performing a value analysis to improve results. Replace any missing values in the database's pre-processing of the normalized data using the Z-score. By applying the same standard deviation, normalization is used to lower the variance in dataset ratings. This illustrates how a certain data point deviates by standard deviation from the mean. The patient data is organized in a vector matrix with columns and rows, where 'n' is the number of patients and 'is' is the number of characteristics. It is used when utilizing several scales or measuring units to normalize variables. The Z-score is normalized by deducting the dataset from each data point and dividing the result by the normative variation. To calculate, use the following formula. A dataset's mean and standard deviation are represented by the z-score of a data point, **W** in the dataset: $z = (x - \mu) / \sigma$. It also recommends meta-training utilizing the effective z-score feature normalization.

$$(Z - score)_U = \frac{(U_{th} - \mu)}{\sigma} \tag{1}$$

The above equation states that the difference between the higher threshold "*U^{th}* and the mean of the characteristics and their standard deviation, or", "determines the highest permissible Z-score.

Feature Extraction using Principal Component Analysis (PCA)
By eliminating all or some of the minor principal components and substituting them with a lower-dimensional projection, a technique known as PCA can be used to reduce the number of dimensions that raw feature data represent while maintaining the highest level of variation in the original data. In this instance, a space is transformed into another via orthogonal linear projection. Here is a description of the PCA method.

$$KY = XC \tag{2}$$

With $\mathbf{Y} \in \mathbb{R}^{S\times S}$ contains the anticipated data matrix ***P*** primary ***X*** Elements with ***P*** ≤ ***X*** elements. The crucial step is to find the projection matrix. $\boldsymbol{C} \in \mathbb{R}^{N\times P}$ This procedure equates to figuring out the eigenvectors of the **X**covariance matrix, or, to applying ***X's*** singular value decomposition (SVD) to solve a problem.

$$X = U\Sigma V^T \tag{3}$$

Where $\mathbf{U} \in \mathbb{R}^{S\times S}$ and $\mathbf{V} \in \mathbb{R}^{S\times ST}$ are the matrix orthogonal in the row and column dimensions of ***X***, and Σ represents single values as a diagonal matrix, **λn** for **n** = 0, · · ·, *M* –1, gradually not lying on the diagonal. Matrix for projecting **C** derived from the initial ***P*** columns of ***V*** with

$$V = [V_1, \ldots, V_n] \tag{4}$$

In addition to

$$C = [C_1, \ldots, C_n] \tag{5}$$

Where $\mathbf{V_n} \in \mathbb{R}^{N\times 1}$ the nth valid specific route of ***X***, and $C_n = V_n$ the truth is each of the distinct values in Σ the ***X*** dispersion measurements along the central axis of the area bounded by the rows of D is shown in (3). Accordingly, λ_n^2 becomes a measure of dispersion over the projection of ***X's*** nth principal component. It is believed that a suitable proxy for the variation of the data is the amount of information that a specific part provides for the overall portrayal of the data. Calculating the proportion of cumulative variance explained between the primary components in one strategy.

$$\boldsymbol{R_{cew}} = \frac{\sum_{n=1}^{P} \lambda_n^2}{\sum_{n=1}^{N} \lambda_n^2} \tag{6}$$

The findings indicate that more than 90% of the total variance or data of ***X*** may be preserved by keeping a few key components. The results of the investigation are compared utilizing several different fundamental parts.

Classification using Adaptive Spider Monkey Swarm Optimized Enhanced Deep Neural Network (AMSO-EDNN)
In artificial intelligence and optimization, the Adaptive Spider Monkey Swarm Optimized Enhanced Deep Neural Network (AMSO-EDNN) is a novel strategy. To improve learning and decision-making, it blends the flexibility of spider monkey swarms with cutting-edge DNN. This hybrid system represents an important advancement in AI research and applications since it has the potential to solve complicated issues.

Enhanced Deep Neural Network (EDNN)
An artificial neural network having several layers between the input and output levels is called an EDNN. Stackable neural networks or neural networks with several layers are good analogies for EDNNs. EDNNs, or deep neural networks, are robust networks constructed using the ANN designs.

In place of multidimensional attributes, EDNN was used to link complex terminals. When estimating agricultural productivity, the environment, plant genotype, and other plant-growing variables are tied together. In this instance, maize and rice crops were chosen, their genotypes recorded, and training on the input and output stages of EDNN have several layers that are completely hidden. Figure 10.3 depicts the EDNN model that was used in this experiment. The network's training approach is backpropagation. Nodes on the input level, or first layer are made up of traits or variables. Data is received from the first input layer by the result level, the next layer also known as the invisible layer. The neurons in the output layer hang onto the output vector once the stimulation function is applied. The lowest layer, the output layer, comprises neurons that serve as placeholders for the intended outcomes. Each layer is linked to the layer underneath via weights, bias and an activation function.

Adaptive Spider Monkey Swarm Optimization (ASMSO)

The cooperative and adaptable behavior of spider monkeys serves as the model for the ASMSO algorithm, a naturalistic optimization method. This ground-breaking method resolves complicated optimization issues by fusing the swarm intelligence of a group with the agility and flexibility of spider monkeys in navigating complex surroundings. ASMSO is a potential technique for taking on complex optimization tasks because it uses the monkeys' capacity to modify their tactics in response to environmental changes and the swarm's collective intelligence. An iterative population-based method is the SMO algorithm. It consists of seven essential processes, including population initialization. In the following several subsections, each phase is described.

Initializing the Population

N spider monkeys started as a colony. The beginning population is represented by the d-dimensional vector $\boldsymbol{SMO_{ji}}$ where i = 1, 2,., N. Each SMO is a practical resolution to the consideration problem. Each $\boldsymbol{SMO_{ji}}$ is initialized using equation 7.

$$SMO_{ji} = SMO_{\min j} + \phi \times (SMO_{\max i} - SMO_{\min i}) \quad (7)$$

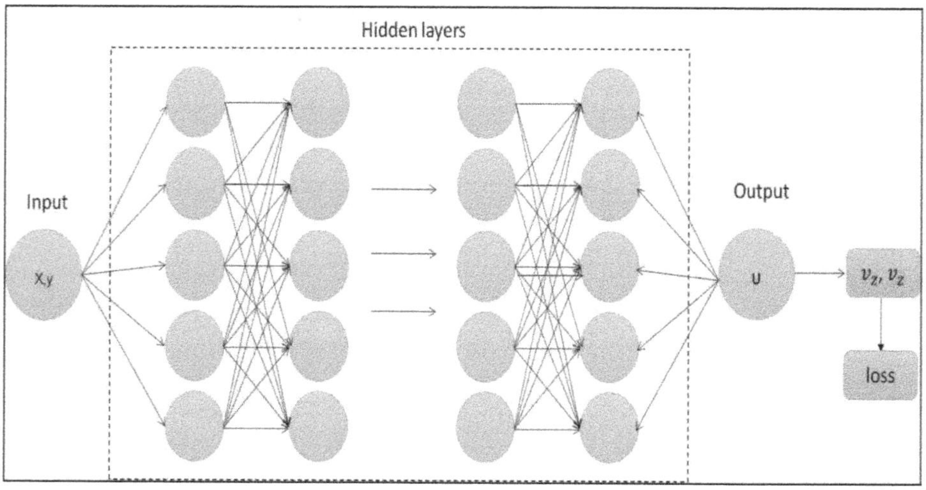

Figure 10.3 Structure of Enhanced Deep Neural Network
Source: Schematic diagram prepared for this research study

$Where \phi \in (0,1)$

Here $SMO_{\min j}$ and $SMO_{\max i}$ show the bottom and higher boundaries with SMO_{ji} in j^{th} orientation.

Local leader phase (LLP)
LLP in light of the experiences of the LL and members of the group, the SMO updates its current location. It compares the fitness of the current and new sites via a meticulous process. j^{th} and the lth limited number changes its positioning by Equation (8).

$$SMO_{new\,ji} = SMO_{ji} + rand[0,1] \times (LL_{li} - SMO_{ji}) + rand[-1,1] \times (SMO_{qi} - SMO_{ji}) \quad (8)$$

LL_{li} Stands for the lth local group leader location in the ith dimension, and SMO_{ji} stands for the jth SMO in that dimension. The qth SMO, SMO_{qi}, is chosen randomly from the lth group so that $q \neq j$ is in the ith extent.

Global leader phase (GLP)
After the LLP is over, the GLP starts. SMO updates their position by utilizing Equation (9) and the knowledge of the GL and local group members.

$$SMO_{new\,ji} = SMO_{ji} + rand[0,1] \times (GL_{i} - SMO_{ji})\ rand[-1,1] \times (SMO_{qi} - SMO_{ji}) \quad (9)$$

Where $i \in \{1, 2, ..., c\}$ signifies a chosen index and GL_i stands for the global leader's position in the ith dimension.

$$q_j = 0.9 \times \frac{fitness}{fitness_{max}} + 0.1 \quad (10)$$

With the support of probability SMO_j adjusts its positions. Estimating the likelihood of a solution based on fitness. There are many other ways to calculate fitness and probability; in this case, q_j was calculated using Eq. (10).

Global leader learning (GLL) phase
Recently, GL relocated using specific greedy strategies. In the current swarm, GL is a suitable response. Additionally, it confirms the GL position has changed and adjusts the global limit count as necessary.

Local leader learning (LLL) phase
The LL changes its position by employing several greedy techniques. It is an appropriate response in the LL group selected for the present swarm. In addition, it determines the LL position has changed and adjusts the local limit count.

Local leader decision (LLD) phase

$$SMO_{new\,ji} = SMO_{ji} + rand[0,1] \times (GL_{i} - SMO_{ji})\ rand[-1,1] \times (SMO_{qi} - GL_{i}) \quad (11)$$

During this phase, a choice is made on the Local Leader position, if it has not been modernized to the local leader limit (LL_{limit}). If there is no change, the location of LL is initialized at random. Eq. (11) may be used to determine the position of LL. Global leader decision (GLD) phase. During this step, a decision is taken on the role of the GL and then little subgroups if it is not modernized to a specific degree, known as the GLL (LL_{limit}). The top limit on the number of subgroups is the maximum number of groups. The LLL technique is used at this stage to select the LL for created subgroups. The LL limit should be D*N with dimension D. The optimal perturbation rate and GL limit are [0.1, 0.9] and [N/2, 2N], respectively. Even though the settings suggested above for the best results, these parameter values are not fixed and may change based on the requirements of the problems. The ASMSO algorithm is depicted in Algorithm 1.

Algorithm 1: Adaptive spider monkey swarm optimization (ASMSO)

Set up each parameter.

Determine fitness

Selecting leaders at all levels

Continue until the extermination requirement is not met.

Create new locations for each member of the group.

$SMO_{new\ ji} = SMO_{ji} + rand[0,1] \times (HK_i - SMO_{ji}) + (q_j) \times (TN_{qi} - TN_{ji})$ here o_j is probability

Between the current location and the recently computed position, use the greedy selection process.

Calculate the likelihood $o_{\square}$f (o_j denotes the probability of jth solution) for each group member,

$$q_j = 0.9 \times \frac{fitness}{fitness_{max}} + 0.1$$

Create new roles for each group member.

$SMO_{new\ ji} = SMO_{ji} + rand[0,1] \times (HK_i - SMO_{ji}) + + (q_j) \times (TN_{qi} - TN_{ji})$

Modernize the roles of local and international leaders. If a local leader's opinion does not change after LL_{limit}t then take the next action.

$(V(0,1) > oq)$

$TN_{new\ ji} = TN_{ji} + rand[0,1] \times (TN_{max\ ji} - TN_{min\ ji})$

Else

$TN_{new\ ji} = TN_{ji} + rand[0,1] \times (TN_{max\ ji} - TN_{min\ ji}) + (TN_{ji} - KK_{li}) \times$

$$\left(\omega_{max} - \frac{iter}{max - iter}(\omega_{max} - \omega_{min})\right)$$

If the global leader ranking does not update after $HK_{L\ it}$ then takes the next action.

If $Global\_Limit\_Count > HK_{Limit}$

Then

$SetGlobal\_Limit\_Count = 0$

If (number of groups <NH)

Then, group the populace accordingly

Otherwise, combine all of the groupings into one giant group.

Update the local leader's position.

Result

A statistical parameter called accuracy evaluates how accurate the forecasts were. It is measured as the proportion of correct forecasts to all predictions by comparing the predictions made by a model to the actual value. After a given number of epochs, the accuracy seems to plateau. In this instance, accuracy appears to reach a table at 40 epochs, when it is 79%. Between 45 and 50 ages, accuracy increases, rising from 80 to 98%. This may result from several things, such as the model arriving at a suitable solution or the availability of an instructive batch of data at that point. The model is doing well on the job, as evidenced by the accuracy of 98% at 50 epochs, as shown in Table 10.2 and Figure 10.3.

$$\text{Accuracy} = \frac{\text{TP+TN}}{\text{TP+TN+FP+FN}} \quad (12)$$

The percentage of cases that are categorized to all occurrences of positive data is known as precision. Between 5 and 10 epochs, there is a considerable increase in accuracy, going from 30 to 45%. Accordingly, the early training epochs appear crucial for accuracy. Precision rises with successive generations, reaching 96% at 50 periods. This shows that the model is quite effective at detecting positive situations and that it is probably attaining a high true positive rate. The accuracy values appear to increase as the number of epochs rises, with no apparent plateaus or declines. This indicates a model has been trained well depicted in Table 10.2 and Figure 10.4.

$$\textbf{precision} = \frac{\textbf{TP}}{\textbf{TP+FP}} \quad (13)$$

The ability of a model to identify every significant sample included in a data collection is referred to as its recall. Recall, sometimes referred to as sensitivity or the proportion of true positives, is a statistical metric that assesses a classification model's capacity to distinguish true positives from false positives when there are many positive cases. As shown in Figure 10.5 recall increases from 27 to 35% between 5 and 10 epochs improving. This suggests that enhancing the model's capability to identify positive instances requires early training epochs. Recall keeps rising with further epochs until it reaches 99% at 50 epochs. This high recall number indicates that the model is very good at collecting almost all positive cases in the dataset.

Table 10.2 Numerical outcomes of accuracy and precision.

No. of epochs	Accuracy (%)	Precision (%)
5	25	30
10	30	45
15	38	57
20	42	68
25	48	71
30	56	78
35	68	80
40	79	88
45	80	92
50	98	96

Source: Based on Primary data from field study

Contrary to precision, recall doesn't seem to peak or decline as the number of epochs rises. The recall values get better as practice more. The model is performing well in terms of recognizing positive cases, and it is probably obtaining a meagre false negative rate, as evidenced by a recall of 99% at 50 epochs, as shown in Table 10.3 and Figure 10.6.

$$\textbf{Recall} = \frac{\textbf{FN}}{\textbf{FN+TP}} \quad (14)$$

The precision and recall components of the F1 score are used to evaluate a model's performance in binary classification tasks. It combines these two criteria into a single value to provide a fair assessment

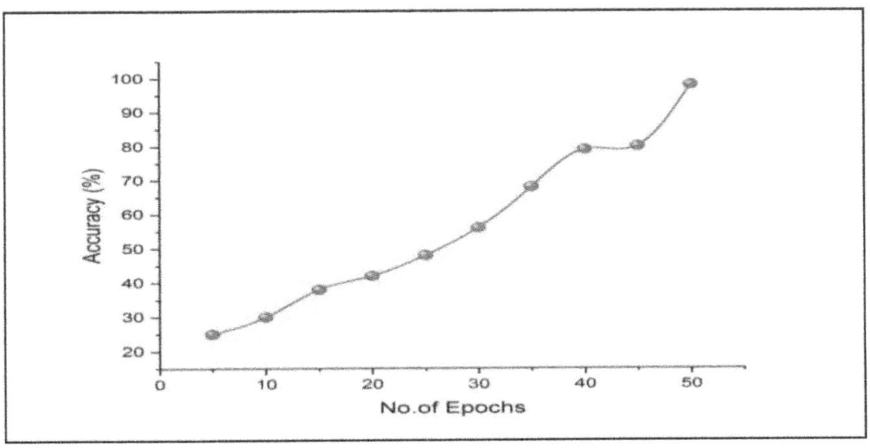

Figure 10.4 Result of accuracy
Source: Based on Primary data from field study

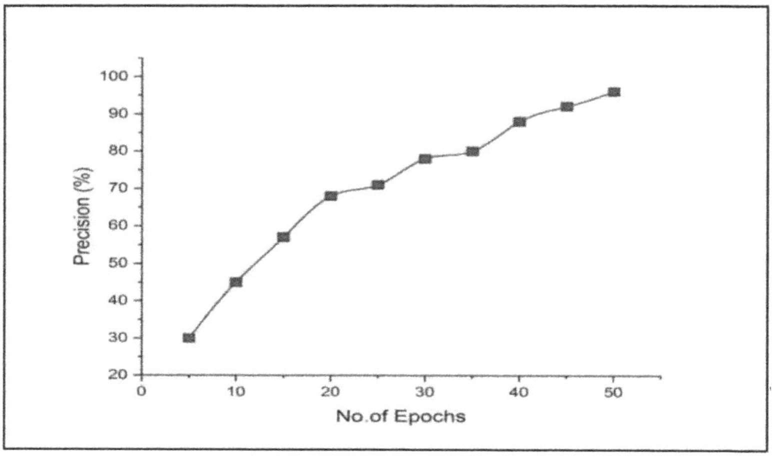

Figure 10.5 Result of precision
Source: Based on Primary data from field study

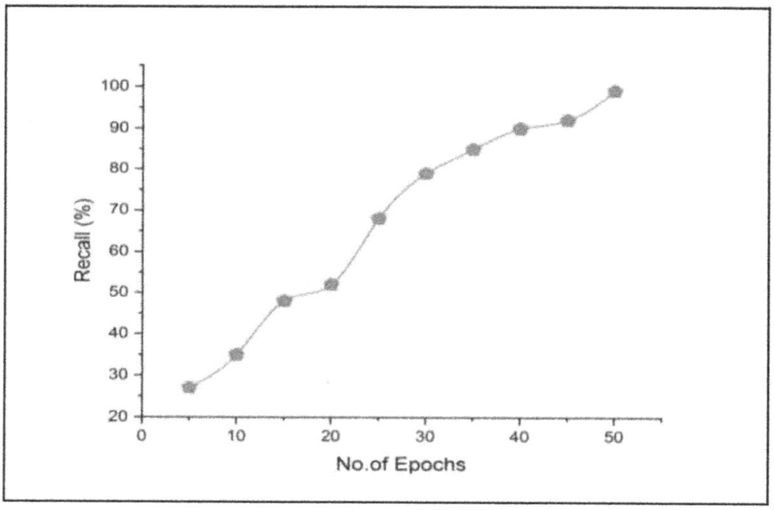

Figure 10.6 Result of recall
Source: Based on Primary data from field study

of the model's accuracy. For the suggested approach, "recall and precision" are combined into a single metric known as the f1-score by computing the harmonic mean. The F1-score begins training at a low value of 17% at five epochs, suggesting subpar overall model performance.

As the number of epochs rises, the F1 score gets better. This shows that when the model is trained more, it becomes more adept at striking a balance between precision and recall. Around 25 epochs, the F1-score exceeds the 50% threshold, indicating that the model is beginning to strike a better balance between precision and recall. This shows that the model is improving its ability to categorize both positive and negative cases. With successive epochs, the F1-score rises slowly until it reaches 95% at 50 epochs. This is a very high F1 score, demonstrating the model's excellent performance in terms of recall and precision. The F1-score doesn't seem to plateau or drop as the number of epochs rises, in contrast to accuracy and memory. With more practice, the F1 score keeps growing in value, as shown in Table 10.3 and Figure 10.7.

$$\mathbf{F1-score} = \frac{\mathbf{(precision)\times(recall)\times 2}}{\mathbf{precision+recall}} \quad (15)$$

Table 10.3 Numerical outcomes of recall and F1 score.

No. of epochs	Recall (%)	F1-score (%)
5	27	17
10	35	20
15	48	25
20	52	30
25	68	49
30	79	52
35	85	65
40	90	75
45	92	89
50	99	95

Source: Based on Primary data from field study

Figure 10.8 and Table 10.4 display the distribution and modification of AP amount, showing the access point amount with the ASMSO-EDNN framework is higher, which is raised by 4.2%. The wheat mite (WM) has a value of 76%, while Sitobion Avenue (SA) has a discount of 79%, Schizaphis germanium (SG) is at 77%, Rhopalosiphumpadi (RP) has a discount of 87%, and Sticky worm (SW), Rice plant hopper (RPH), and Dichocrocispunctiform (DP) all denote values of 71%

The corn borer and the bollworm are mentioned as examples of agricultural pests (AP). Figures 10.9 (A) and 10.9 (B), which use the AP information as their test set, show the precision-recall curves derived from multiple target identification methods. The proposed ASMSO-EDNN method outperforms existing target detection techniques in terms of accuracy and recall rate. It forms a developing law that exhibits comparable proportions and alterations to the MPD information.

The results of the MPD and AP collections provide useful information for detecting and naming crop hazards in the wild, demonstrating that the suggested CNN-based pest concentrate detection technique

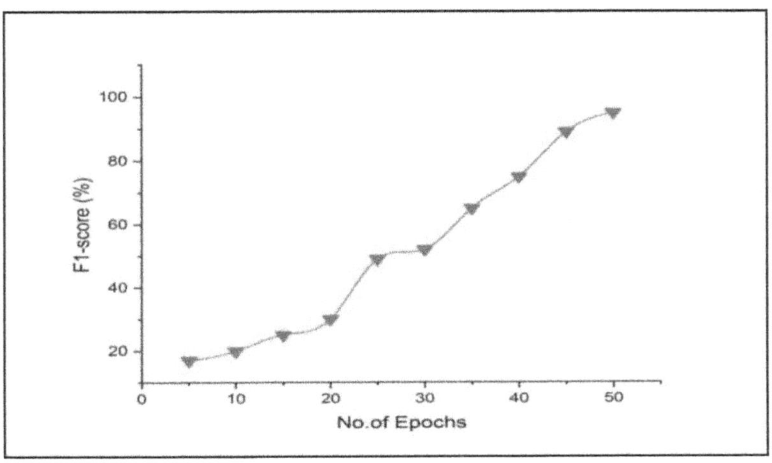

Figure 10.7 Result of F1-score
Source: Based on Primary data from field study

Table 10.4 Numerical outcomes for AP value.

Name of the pest	AP value
WM	76
SA	79
SG	77
RP	87
SW	73
RPH	77
DP	77

Source: Based on Primary data from field study

has a presenting effect and lowers the number of incorrect identifications and failed identifications. The method provides a high level of detecting precision. The results demonstrate that the ASMSO-EDNN technique for deep learning-based pest identification may be used by the platform for agricultural information management and administration.

The study's findings and recommendations show that the suggested strategy increases pest detection accuracy while

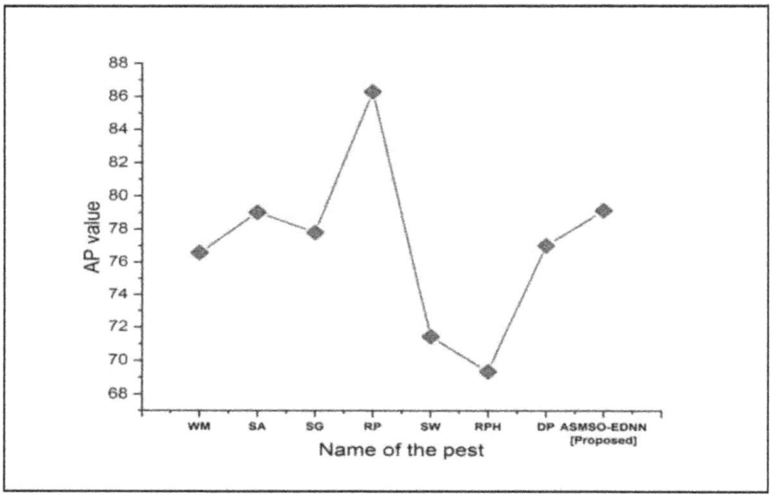

Figure 10.8 AP value
Source: Based on Primary data from field study

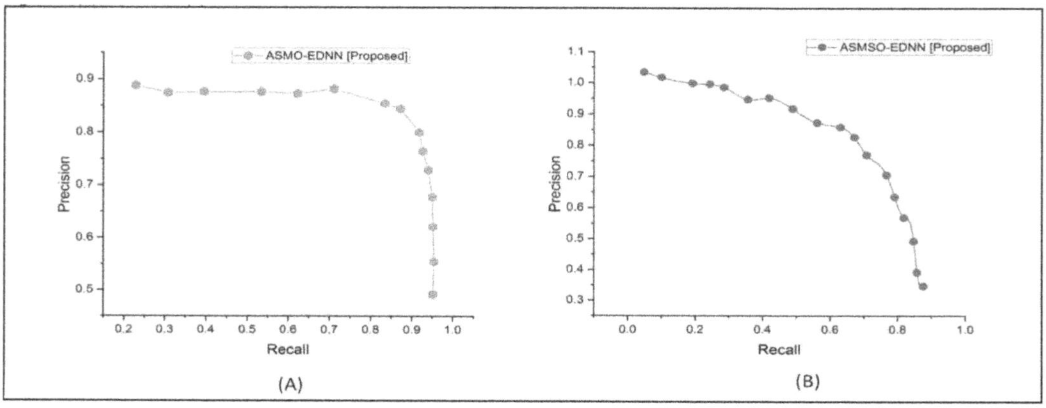

Figure 10.9 (A) Recall and (B) precision curves
Source: Based on Primary data from field study

lowering false positives and false negatives. The authors cover the application of blockchain technology to pest control, highlighting the significance of pattern recognition in the processing of agricultural data. They contend this strategy improves the long-term survival of rural communities and raises knowledge of farming data management.

Conclusion

This work offers a unique, deep learning-based insect target detection system called ASMSO-EDNN that is suited for rural companies, considering the historical transition from organic to inorganic farming in agriculture. It uses principal component analysis (PCA) and picture target detection to prepare the data while overcoming financial limitations. The findings demonstrate higher precision-recall curves, decreased false detections, and improved agricultural data processing awareness, highlighting the possibility for SD and governance in rural areas using cutting-edge technology.

Limitations and Future Scope

There are several restrictions when deploying intelligent management systems in rural regions for corporate profitability and SD. Limited access to cutting-edge technology, a lack of digital infrastructure, technical knowledge in rural areas, high startup costs, and the requirement for solutions customized to different rural contexts are a few problems. Realizing the full potential of such systems in rural development will depend on overcoming these challenges. Utilizing intelligent organizational structures that optimize resource allocation, encourage eco-friendly behaviors and increase corporate profitability, supporting economic growth and environmental stewardship, the future offers immense potential for sustainable advancement in rural areas.

References

Acuti, D., Bellucci, M., and Manetti, G. (2020). Company disclosures concerning the resilience of cities from the SDGs (SDGs) perspective. *Cities*, 99, 102608.

Charania, I., and Li, X. (2020). Smart farming: agriculture's shift from a labour-intensive to technology-native industry. *Internet of Things*, 9, 100142.

Ciliberto, C., Szopik-Depczyńska, K., Tarczyńska-Łuniewska, M., Ruggieri, A., and Ioppolo, G. (2021). Enabling the CE transition: a sustainable lean manufacturing recipe for Industry 4.0. *Business Strategy and the Environment*, 30(7), 3255–3272.

Clark, M. (2018). Teleworking in the Countryside. Routledge. https://doi.org/10.4324/9781315189697.

Fina, B., Auer, H., and Friedl, W. (2019). Profitability of PV sharing in energy communities: use cases for different settlement patterns. *Energy*, 189, 116148.

Gupta, S., Prathipati, B., Dangayach, G. S., Rao, P. N., and Jagtap, S. (2022). The development of a structural model for the adoption of Industry 4.0 enables sustainable operations for operational excellence. *Sustainability*, 14(17), 11103.

Han, J., Heshmati, A., and Rashidghalam, M. (2020). CE business models with a focus on servitization. *Sustainability*, 12(21), 8799.

Harris, M., Yao, H., Tariq, G., Malik, A., and Javaid, H. M. (2019). Intellectual capital performance and profitability of banks: evidence from Pakistan. *Journal of Risk and Financial Management*, 12(2), 56.

Khan, I. S., Ahmad, M. O., and Majava, J. (2021). Industry 4.0 and SD: a systematic mapping of triple bottom line, CE and sustainable business models perspectives. *Journal of Cleaner Production*, 297, 126655.

Nayal, K., Kumar, S., Raut, R. D., Queiroz, M. M., Priyadarshinee, P., and Narkhede, B. E. (2022). Supply chain firm performance in CE and digital era to achieve SD goals. *Business Strategy and the Environment*, 31(3), 1058–1073.

Nayal, K., Raut, R. D., Narkhede, B. E., Priyadarshinee, P., Panchal, G. B., and Gedam, V. V. (2021). Antecedents for blockchain technology-enabled sustainable agriculture supply chain. *Annals of Operations Research*, 7(4) 1–45.

Niu, Y., Ying, L., Yang, J., Bao, M., and Sivaparthipan, C. B. (2021). Organizational business intelligence and decision-making using big data analytics. *Information Processing and Management*, 58(6), 102725.

OECD (2018). Bridging the Rural Digital Divide. OECD Publishing, Paris, OECD Digital Economy Papers, No. 265 https://dx.doi.org/10.1787/852bd3b9-en.

Rafiq, M., Zhang, X., Yuan, J., Naz, S., and Maqbool, S. (2020). Impact of a balanced scorecard as a strategic management system tool to improve SD: measuring the mediation of organizational performance through PLS-smart. *Sustainability*, 12(4), 1365.

Tien, N. H., Minh, H. T. T., Mai, N. P., and Thuc, T. D. (2020). Social entrepreneurship and corporate SD. evidence from Vietnam. *Cogent Business and Management*, 7(1), 1–17.

Turban, E. (2022). Decision Support and Business Intelligence Systems. (3rd Edn), Pearson Education India, pp. 68–77.

Wang, R., Liu, L., Xie, C., Yang, P., Li, R., and Zhou, M. (2021). Agripest: a large-scale domain-specific benchmark dataset for practical agricultural pest detection in the wild. *Sensors*, 21(5), 1601.

White, M. A. (2023). Sustainability: i know it when i see it. *Ecological Economics*, 86, 213–217.

Xu, J., and Liu, F. (2020). The impact of intellectual capital on firm performance: a modified and extended VAIC model. *Journal of Competitiveness*, 11(1), 87–99.

11 Sustainable circular economy through waste to wealth initiative: a study in smart city of Bhubaneswar

Sasmita Mohapatra[1,a] *and Mihir Ranjan Nayak*[2,b]

[1]Assistant Professor, Biju Patnaik Institute of Information Technology and Management Studies (BIITM), Bhubaneswar, Odisha, India

[2]Professor, Biju Patnaik Institute of Information Technology and Management Studies (BIITM), Bhubaneswar, Odisha, India

Abstract

The current study focuses on municipal solid waste management (MSWM) practices in the city of Bhubaneswar and examines the relationship between city residents' occupations and the awareness level of the individuals for the circular economy through the waste wealth initiative. It primarily examines the five key components of waste management: collection, segregation, transportation, recycling, and disposal. The study considers Ajzen's theory of planned behavior (TPB) which mentions that the attitude of individuals toward behavior, subjective norm, and perceived behavioral control, has an impact on calculating the behavioral intent and real behavior of individuals when participating in decisions are intentional and under individual control (Ajzen, 1991). The study also aims to discover the perception of people towards solid waste management practices and the awareness among the people of the smart city. It also explores the possibilities for various policy changes and possible approaches to make the current municipal solid waste management process more effective through a quantitative study. SPSS tool was used to complete the analysis and Microsoft Excel was used to cleanse and harness the data. The findings of the study aim to guide policymakers, urban planners, and sustainability stakeholders in formulating more sustainable and efficient waste management solutions that contribute to the circular economy and strengthen sustainability in smart cities in India.

Keywords: Bhubaneswar, circular economy, smart city, smart city, sustainability, waste to wealth

Introduction

In pace, with the rapid population growth and massive urbanization solid waste generation and its scientific management have been a major concern in India. The smart city of Bhubaneswar, Odisha is also struggling with this new challenge of effective solid waste management. Managing the generation, storing, gathering, transmission, processing, and clearance of solid wastes in an approach that fulfills the quality parameters, economics, well-being, preservation, and other environmental concerns are acknowledged as solid waste management (Tchobanoglous et al., 1993). Improper solid waste management is one of the primary sources of Pollution in a city (Böhmar, 1995). However, the concept of "Waste to Wealth" offers a viable remedy for this crucial problem, opening the door to sustainability and circular economy. The study captures the spirit of this revolutionary methodology. To support the city's sustainability goals and keep them in line with national and international sustainability goals, this study intends to investigate the possibility of turning trash into

[a]sasmitamohapatra@biitm.ac.in, [b]nayak.mihirranjan@gmail.com

DOI: 10.1201/9781003564447-11

useful resources. Bhubaneswar produces a substantial amount of solid waste. As per the figures of Bhubaneswar Municipal Corporation (BMC) and other agencies, the city generates around 520 metric tons of waste every single day. This waste comprises household, commercial as well as industrial waste. It is also calculated that in Bhubaneswar, each person produces about 500 grams of garbage every day on average. To sum it up together the city generates approximately 182,500 metric tons of municipal solid waste (MSW) annually as per the BMC website.

The Problem Statement

The city of Bhubaneswar is generating a lot of solid waste on a day-to-day basis and BMC is collecting the solid waste and processing it further. However, there seems to be a lack of awareness among city residents related to waste to wealth generation. People are carelessly mixing wet and dry waste which is causing solid waste segregation a challenge. People are not aware of the reusability of plastic waste and the use of single-use plastic is rampant. The collection process involves segregation but due to lack of awareness, the segregation at source is not always adhered to by the city residents. There is no incentive for city residents who have good awareness levels and follow instructions at the same time the penalty clauses are not so stringent.

Research Objectives

The objectives of this research are as follows:

1. To determine the level of awareness of Bhubaneswar city residents regarding the circular economy through solid waste management practices in different occupational groups.
2. To examine if there is a significant difference between various occupational groups of city residents related to awareness of circular economy through solid waste management initiatives.

Research Questions

As per the research objectives, the following research questions are formulated

What is the level of awareness related to the circular economy through solid waste management initiatives in various occupational groups in the city of Bhubaneswar?

Is there any significant difference between various occupational groups of Bhubaneswar city residents related to the circular economy through solid waste management initiatives?

Research Hypotheses

Considering the research questions the following null and alternative hypothesis is formulated.

H0: There is significant difference between various occupational groups of Bhubaneswar city residents related to awareness about the circular economy through solid waste management initiatives.

H1: There is no significant difference between various occupational groups of Bhubaneswar city residents related to awareness about the circular economy through solid waste management initiatives.

Theoretical Framework

This current research is grounded on the theory of planned behavior (TPB) by Ajzen (1991). According to the (TPB), an individual's behavior is influenced by their communicative, socio economic, and control beliefs. It also suggests that individuals' behavior is influenced by their defiance, individual standards, and apparent control when taking voluntary decisions.

Literature Review

BMC's Public Health Department is in charge of gathering, moving, and disposing of the solid waste produced in the Smart City of Bhubaneswar. Regarding solid waste management, BMC's primary priority areas include trash segregation at the point of creation, waste collection from home to house, road cleaning, night cleaning in marketplaces and crowded locations, and waste transportation to the appropriate disposal site. Unpredictable variations in waste quantity and type also put policymakers under pressure to act swiftly. However, practically all of the solid trash is thrown throughout the city carelessly (Mohanty et al., 2022). Approximately 400 tons of solid trashes are produced daily on average (500 gm/day/capita). Out of this 8.8% constitute leather and plastic, 1.25% constitutes metal and glass, 28.15% inert material, and 61.8% biodegradables make up the typical solid trash (Sethy et al., 2019). Ten of the 67 wards of Bhubaneswar are managed by the BMC, while 57 of the 67 wards receive sanitation services from private organizations. Households, hotels, markets, temples, and street sweeping are among the places that produce solid waste in urban areas. Inefficient garbage management causes unattractive conditions on the streets and the spread of illness. As the population is growing in urban areas solid waste is also substantially increasing creating health hazards and in turn, polluting the groundwater (Babu, 2013). Researchers have done studies related to different kinds of solid waste generated in urban areas and how to effectively deal with it in a sustainable way. A study completed emphasizes green technologies that convert waste into valuable products like electricity, biofuel, and biopolymers without harming the environment. The comprehensive coverage includes case studies from the US and Japan, offering insights into advanced bioremediation methods (Rai and Saraswat, 2023). On similar endeavour done for end-of-life tires explores modelling approaches, emphasizing cost-effectiveness and enhanced durability. Despite challenges, the research advocates for further innovation, highlighting the potential of scrap tires to create a more sustainable and resilient built environment (Shariati et al., 2024). There exists a relationship between waste generation, greenhouse gas emission, economic growth, and circular economy framework. Research conducted in Switzerland indicates that both municipal waste and economic growth have a substantial effect on greenhouse gas (GHG) emissions in both the short and long term. Causality analyses show that there are unidirectional causal links between municipal solid waste and economic growth to GHG emissions, as well as bidirectional causality between municipal waste and economic growth. The study highlights critical factors in the areas of "policy drivers" and "environment and health," stressing the significance of mission-oriented policy strategies in effectively reducing GHG emissions and promoting the circular economy (Magazzino and Falcone, 2022). Municipal solid waste has a great potential to create organic compost which can completely replace the chemical fertilizers that are used in cultivation. A study done in Ethiopia explores the potential of composting organic waste and human excreta to reduce landfill waste, enhance soil fertility, and boost agricultural yields. It measures household solid waste (HSW) and organic waste from various sources during wet and dry seasons (Obsa et al., 2022). Solid waste generated in urban areas is a mix of Biodegradable and non–Non-biodegradable components. Both of these have the potential to contribute to the modern age circular economy. The circular economy (CE) model, focusing on resource optimization and waste minimization, offers a viable solution for waste management. This study discovers the effective use of both biodegradable and non-biodegradable parts of solid waste within integrated refinery platforms. The biodegradable fraction can be processed using advanced biological methods, generating bioenergy such as bio hydrogen, bio methane, and valued substances like ethanol and methanol (Velvizhi et al., 2020). To enable similar people-centric and people-driven waste management initiatives some incentives for good waste management, such as best lane/street awards, should be provided to both workers and the general public. Similarly, a penalty system must also be put in place to fine the community and the workers for not changing their attitudes and behaviors (Sahu and Mishra, 2023). To make massive initiatives like Waste to Wealth greater public awareness and inclusion are required so that people will consider this as their individual initiative by organizing public awareness campaigns related to hazards of pollution and air pollution caused by municipal solid waste (BMC website) (BMC, 2024).

Data and Research Methodology

Primary data was collected from the respondents through a structured questionnaire. A set of questions with demographic information like age group, gender, marital status, and occupation were prepared. These questions had 5-point Likert scale-type answers and the same was collected online by survey method through Google Forms. Table 11.1.

Table 11.1 Identified Variables (using computed variable technique) for the Analysis along with their Category Names to be used in SPSS for Analysis.

Variable category	Subject area
Awareness	Pay for solid waste management segregation
	Health and well-being related to waste management
	Waste to wealth at home
	Single-use plastic awareness

Source: Authors' own compilation using SPPS tool

A total of 159 responses were collected from Bhubaneswar city residents which included male, female, married, and unmarried individuals. The responses were collected through a Google form through different digital channels like email and other social media platforms. The questions were intended for an organized survey on solid waste management and BMC's drive for waste to wealth. In addition, the respondents were separated into different groups according to their gender, age, marital status, and occupation. After the responses were recorded three of the responses were rejected due to incompleteness of information and finally 156 responses were considered for the final analysis. SPSS tool was used to complete the analysis and Microsoft Excel was used to cleanse and harness the data.

Data Analysis and Interpretation

Demographics of the Respondents

Out of 156 respondents, there were 62 respondents were married, whereas 94 respondents were unmarried. The number of male respondents was 87 (approximately 56%) and female respondents were 69 (approximately 44%). The count of respondents as per the age group are 18 to 24 years (44), 25 to 34 years (24), 35 to 44 years (21)45 to 54 years (28), and above 55 years (39). Table 11.2.

Descriptive statistics were performed on the prepared dataset to determine the count, mean, standard deviation, and skewness, assessing the dataset's suitability for further analysis. This analysis was conducted using the SPSS tool, and the results are detailed below. The descriptive statistics provide the mean for each variable and indicate the skewness of the data. Since the skewness values fall within the range of (+/-) 2, the data are deemed appropriate for further analysis.

The waste-to-wealth management activities of BMC with regard to solid waste management are evaluated in the current study. The purpose is to identify how it can

Table 11.2 Analysis of variance with for Awareness.

Model		Sum of squares	df	Mean square	F	Sig.
1	Regression	5.713	1	5.713	11.197	.001[a]
	residual	79.089	155	.510		
	Total	84.802	156			

a. Predictors: (Constant), Occupation

b. Dependent variable: Awareness

Source: Authors' own compilation using SPPS tool

Table 11.3 Analysis of Variance with for Awareness.

Model	Unstandardized coefficients		Standardized coefficients	t	Sig.	95% Confidence interval for B	
	B	Std. Error	Beta			Lower bound	Upper bound
Constant	3.822	.137		27.968	.000	3.552	4.091
Occupation	-.169	.050	-.260	-3.346	.001	-.268	-.069

Source: Authors' own compilation using SPPS tool

add value to the circular economy for the Smart City of Bhubaneswar.

As the identified variables are more in number they were grouped into different categories using the computed variable technique in SPSS. The variables related to awareness are considered for further analysis. The below table represents the categories and variables mapped in tabular form used in SPSS.

The effect of occupation was analyzed with the first with respect to the four computed variables categories (Awareness, Behavioral, Segregate and Authority). Multiple regression analysis was done to find out which independent variable has more impact on the dependent variable awareness.

The dependent variable (awareness) was regressed on the independent variable occupation. These independent variables significantly predict the occupation, $F (1,155) = 11.197$, $p > .001$. The significance value is .001. These indicate that occupation does have a significant impact on awareness. However, it does not indicate that all four variables are significant. To examine it further, the coefficient table needs to be analyzed for all the variables for awareness. Table 11.3.

To determine if all the variables included in awareness have a positive impact or only a few have. To determine this further it is analyzed by one-way ANOVA which of the variables in awareness have a significant impact on occupation. Table 11.4.

The results indicate that the Health and Well-being related to waste management are significant between the groups. This signifies that the question "Do you agree that your community's cleanliness affects the overall health and well-being of residents?" is significant among awareness variables with respect to the occupation of the respondents. Hence H0 is accepted. This means occupation has a positive impact on the awareness of the waste-to-wealth initiative of the city residents.

Discussion

The conventional method of burning or disposing of wastes results in environmental damage in addition to ignoring the possible

Table 11.4 ANOVA for Different Variables for Awareness.

	Sum of squares	df	Mean square	F	Sig.
Between groups	8.997	3	2.999	1.580	.197
Within groups	290.430	153	1.898		
Total	299.427	156			
Between groups	23.819	3	7.940	4.683	.004
Within groups	259.429	153	1.696		
Total	283.248	156			
Between groups	5.413	3	1.804	1.253	.293
Within groups	220.306	153	1.440		
Total	225.720	156			
Between groups	4.689	3	1.563	.881	.453
Within groups	271.540	153	1.775		
Total	276.229	156			

Source: Authors' own compilation using SPPS tool

benefits that could be obtained from them. Challenged by the 'Waste to Wealth' notion, garbage should be seen as a resource rather than an issue. Solid waste can be processed utilizing the 3 R Method (Reduce, Recycle, and Reuse) to make it suitable for use in another way. However, the awareness about the implications of improper way of solid waste management is at different levels as far as the occupation level of the individuals is taken into consideration. Occupation is mostly related to the education and awareness level of the city residents. The intention, attitude, awareness, as well as perceptibility of the initiatives by the local authority, also seem to have different replications among different occupational groups in the city. The findings of the study disclose that occupation and awareness have a significant impact on the waste-to-wealth initiative. While considering the major factor in different occupation groups the awareness about health and well-being of residents is a significant factor among different occupation groups.

Implications for Research, Community, and Policy

Environmental policymakers need to focus and chalk out the required control and recycling plan so that it can help the circular economy by utilizing waste for wealth generation. Although coined by the Bhubaneswar Municipal Corporation this drive needs to be people with each and every individual is responsible for the circular economy of the city through waste to wealth management.

Incentivizing good citizens who contribute to this massive drive would also help in creating a sustainable circular economy with people's participation. The Smart City of Bhubaneswar is a city of denizens which opens up a scope of E-waste in the city. Separate bins for electronic wastes like old mobiles, charges, computer peripherals, electrical waste, etc., and awareness programs on the same can be the future scope of this study. Scientific waste management of e-waste can boost the waste-to-wealth initiative of the city.

Summary and Conclusion

The Bhubaneswar Municipal Corporation (BMC) has taken the lead in introducing environment-friendly methods of managing solid waste. The 'Waste to Energy' project is one such endeavor that seeks to turn waste into useful resources. This will not only address the waste management issues but will also cater to the circular economy of the city as well as the sustainability aspect. The city has a population of different age groups and occupations, and the opinions and the effectiveness of the drive were analyzed considering their perception and participation in the initiative. Awareness community's cleanliness affects the overall health and well-being of residents is one of the most significant factors among the city residents.

References

Ajzen, I. (1991). The theory of planned behavior. *Organizational Behavior and Human Decision Processes*, 50(2), 179–211.

Babu, Neela Victor. "A study on selection of landfill sites for solid waste management in the environs of greater Visakhapatnam municipal corporation, Andhra Pradesh: a spatial technology approach." (2013).

BMC (2024). https://www.bmc.gov.in/programs/swachh-bhubaneswar-abhiyaan [Last accessed on 20-Jan-2024].

Böhmar, G. (1995). Solid waste and the hierarchy in solid waste management system. *Diphomarbeit van der WirtschaftSuniversität Wien*, 105.

Magazzino, C., and Falcone, P. M. (2022). Assessing the relationship among waste generation, wealth, and GHG emissions in Switzerland: Some policy proposals for the optimization of the municipal solid waste in a circular economy perspective. *Journal of Cleaner Production*, 351, 131555.

Mohanty, S., Mishra, S., and Mohanty, A. (2022). Municipality solid waste management. a case study of smart city Bhubaneswar, Odisha. *Journal of Environmental Management and Tourism*, 13(5), 1361–1373.

Obsa, O., Tadesse, M., Kim, D. G., Asaye, Z., Yimer, F., Gebrehiwot, M., et al. (2022). Organic waste generation and its valorization potential through composting in Shashemene, southern Ethiopia. *Sustainability*, 14(6), 3660.

Rai, J. P. N., and Saraswat, S. (2023). Green Technologies for Waste Management: A Wealth from Waste Approach. CRC Press.

Sahu, C., and Mishra, S. (2023). Public perception on municipal solid waste management: a case of India. *International Journal of Environmental Science and Technology*, 20(4), 4189–4202.

Sethy, K. M., Harichandan, A., Pradhan, S., Ghosh, M., and Patra, S. H. (2019). Threat to ambient air quality due to poor municipal solid waste management: a case study in and arround vuasuni dumping site, Bhubaneswar, Odisha, India. *Journal of Environment and Earth Science*, 9(2), 7–16.

Shariati, M., Afrazi, M., Kamyab, H., Rouhanifar, S., Toghroli, E., Safa, M., et al. (2024). A state of the art review on geotechnical reinforcement with end life tires. *Global Journal of Environmental Science and Management*, 10(1), 385–404.

Tchodanoglous, George, Hilary Theisen, and Samuel Vigil. *Integrated solid waste management: engineering principles and management issues*. McGraw-Hill, 1993.

Velvizhi, G., Shanthakumar, S., Das, B., Pugazhendhi, A., Priya, T. S., Ashok, B., et al. (2020). Biodegradable and non-biodegradable fraction of municipal solid waste for multifaceted applications through a closed loop integrated refinery platform: Paving a path towards circular economy. *Science of the Total Environment*, 731, 138049.

12 Urban sustainability: a multi-criteria analysis of AMRUT cities in Odisha

Koustubh Kanti Ray[1,a], *Anupriya Panda*[2,b], *Joydeep Biswas*[3,c] *and Mohar Banerjee Biswas*[4,d]

[1]Professor, School of Management, KIIT Deemed to be University, Bhubaneswar, India

[2]Research Scholar, School of Management, KIIT Deemed to be University, Bhubaneswar, India

[3]Associate Professor, School of Management, KIIT Deemed to be University, Bhubaneswar, India

[4]Assistant Professor, School of Management, KIIT Deemed to be University, Bhubaneswar, India

Abstract

This paper explores the urbanization phenomenon in India, focusing on the Atal Mission for Rejuvenation and Urban Transformation (AMRUT) launched in 2015. The study employs the analytic hierarchy process (AHP) to evaluate the critical factors influencing urban transformation within the AMRUT framework. There are nine cities in Odisha coming under the AMRUT Project. We have selected only three cities, namely Bhubaneswar, Cuttack, and Rourkela for the present work. A stratified random sampling methodology was applied, surveying 320 respondents in Bhubaneswar, Cuttack, and Rourkela cities. Factors identified through a comprehensive literature review were categorized into five dimensions: core resources, supporting resources, city enhancement factors, city management factors, and broad city factors. Consistency tests were conducted for each dimension, revealing strong consensus among decision-makers. The criteria weights for each factor were determined, highlighting the relative importance within each dimension. The study offers policy recommendations, aiding urban policymakers in identifying critical factors, conducting service gap analyses, ranking cities based on project outcomes, redesigning projects, and formulating effective policies. The societal relevance of the study extends to providing insights into societal transformation, supporting informed decision-making, identifying social requirements and gaps, and evaluating project sustainability from a socio-economic perspective. The findings contribute to a clear understanding of urban development strategies, offering practical implications for policymakers, researchers, and the public.

Keywords: Analytic hierarchy process, amrut, effectiveness, efficiency, mcdm, sustainability

Introduction

Globalization, a process characterized by the increasing interconnectedness and interdependence of the world's economies, cultures, and populations, has significantly influenced urbanization trends across the globe. The phenomenon of globalization fosters urbanization by facilitating the flow of capital, labor, and technology into cities, thereby transforming them into epicenters of economic activity and cultural exchange. As a result, urban areas attract a diverse array of people seeking better employment opportunities, education, and living standards, which accelerates urban growth. The synergy between urbanization and economic growth creates a virtuous cycle where improved economic conditions lead to further urban development. India, one of the fastest urbanizing countries, is rapidly transforming, with millions migrating to cities in search of better opportunities.

[a]drkkray@gmail.com, [b]pandaanupriya@gmail.com, [c]joydeep.biswas@ksom.ac.in, [d]mohar.biswas@ksom.ac.in

DOI: 10.1201/9781003564447-12

During the Eleventh Plan Period, India has demonstrated remarkable economic growth, positioning itself among the world's fastest-growing economies with a growth rate exceeding 8% (Kundu, D. 2015). An Urbanizing India is not just a demographic shift. It is based on three broad factors: (i) Natural growth of population, (ii) rural-urban migration, and (iii) reclassification of rural areas as urban, (Bhagat, 2012).

However, in the 1990s, i.e. when the Indian government opened its economy, there was a rise in rapid economic growth of the country. However, this economic growth was more of urban growth, which led to the rapid migration of rural population to urban areas. In the year 1951, only five cities had a population of more than one million, which increased to 53 cities in 2011 and it was estimated that by 2031, it will be 70 cities in the country with a million population. Similarly, three cities had more than 10 million populations in the year 2011, and it was estimated that six cities will have 10 million populations by 2031. Thus, it was estimated that the total urban population by the year 2030 will be 610 million, which will account for 40% of the total population of the country (Mohan and Kandya, 2015). Many cities in the country, are becoming extremely crowded, and out of the 20 most densely populated cities in the world, five will be from India (Confederation of Indian Industry, 2010).

Urbanization, a process driven by globalization, holds both positive and negative consequences for city sustainability. Economically, urbanization acts as a catalyst for growth by concentrating industries, services, and innovation in cities, thus stimulating productivity and generating job opportunities. Additionally, urban areas typically boast superior infrastructure and access to essential services, enhancing the overall quality of life for residents. However, this rapid urban growth also brings environmental challenges, such as pollution and resource depletion, alongside social issues like overcrowding and socioeconomic inequality. Achieving sustainable urban development requires a balanced approach that leverages the economic benefits of urbanization while mitigating its adverse effects. This entails implementing green initiatives, equitable resource distribution, and resilient infrastructure to ensure cities remain sustainable, resilient, and inclusive environments for all inhabitants.

Urbanization and urban planning need to act cohesively to address the progressive concentration of the population in urban units, (US Conference of Mayors 2014). Concurrently, urban planning emerges as a critical facet, encompassing the strategic composition of urban spaces to achieve sustainable, efficient, and equitable development. It involves careful planning on land use, infrastructure development, transportation networks, housing provisions, and environmental preservation, all aimed at cultivating livable and resilient urban environments. In this context, the Atal Mission for Rejuvenation and Urban Transformation (AMRUT) emerges as a pivotal initiative by the Indian government, to balance urban growth with sustainability. AMRUT aims to provide basic urban infrastructure, improve service delivery, and promote sustainable development in cities, thereby addressing both the benefits and drawbacks of rapid urbanization.

The AMRUT launched in 2015 covers 500 cities nationally with an investment of INR 50,000 crore. AMRUT represents a strategic evolution from its predecessor JNNURM that involves direct participation by citizens and ULBs (urban local bodies) for their cities' improvement. The respective states combine the city development plans and submit them to the Government of India (GoI) as a State Annual Action Plan. AMRUT plays a crucial role in India's urban development roadmap by weaving together these diverse programs into a cohesive strategy. It tackles core infrastructure needs, strengthens smaller and medium-sized cities, complements existing initiatives, and creates fertile ground for additional programs to thrive. AMRUT follows the bottom-up approach and acts as a vital thread that binds these separate initiatives, contributing to a more holistic and sustainable vision for India's urban future. AMRUT is an ongoing mission, with a planning horizon extending up to 2031, indicating a long-term commitment to urban transformation and, thereby making AMRUT more direct, transparent, and relevant.

Ensuring universal access to basic services such as water supply, sewerage, and urban transport, and enhancing urban amenities to elevate the quality of life, particularly for marginalized populations, stands as a paramount national priority. A study commissioned by the High-Powered Expert Committee (HPEC) in 2011 projected an investment requirement of approximately Rs. 39.2 lakh crore over a 20-year period, based on 2009-10 price levels, to develop urban infrastructure. This allocation encompassed Rs. 17.3 lakh crore for urban road networks and Rs. 8 lakh crore for essential services including water supply, sewerage, solid waste management, and stormwater drainage systems. Moreover, an estimated Rs. 19.9 lakh crore was earmarked for the operation and maintenance (O&M) of infrastructure facilities. The strategic focus of infrastructure development must closely align with the genuine needs of the populace, particularly in ensuring universal access to tap water and sanitary facilities for every household. This imperative was underscored by the President of India in addresses to the Joint Sessions of Parliament in June 2014 and February 2015. Accordingly, the AMRUT has been formulated to address these overarching objectives:

- Ensure that each household has access to a tap providing a consistent water supply and a drainage connection.
- Enhance the aesthetic appeal of metropolises by establishing green spaces and maintaining open green zones such as parks.
- Mitigating pollution through the promotion of public transportation or the establishment of infrastructure for non-motorized modes of transport, such as walking and cycling, represents a priority objective. These initiatives are highly esteemed by citizens, notably women, and the Ministry of Urban Development (MoUD) has instituted service level benchmarks (SLBs) to assess advancements towards these goals.

In Odisha, the AMRUT has made significant steps in addressing the unique urban challenges faced by the state. Focusing on nine key cities, the mission has invested a significant INR 1,164 crore, delivering impactful initiatives like constructing over 4.4 lakh individual household toilets for improved sanitation and building 160+ km of stormwater drainage channels for enhanced resilience. Notably, all nine AMRUT cities now boast 100% access to clean water through household tap connections. The study chose three cities out of nine AMRUT cities as the sample study because three cities are regarded as smart and metropolitan cities. AMRUT 2.0 launched in 2022, extends the mission's focus on water supply reforms, city beautification, and climate-resilient infrastructure.

Literature Review

This section reviews the literature on AMRUT's impact on urban sustainability and strategic planning, drawing from various studies and reports. AMRUT's primary objective is to ensure every household has access to a tap with an assured water supply and a sewerage connection. The mission also seeks to increase the amenity value of cities by developing greenery and well-maintained open spaces and reducing pollution by switching to public transport or constructing facilities for non-motorized transport (Bhandari, 2016). The mission targets 500 cities, with a total outlay of Rs. 50,000 crore, focusing on infrastructure improvement, capacity building, and policy reforms. Several studies highlight AMRUT's impact on urban infrastructure. According to Rao and Patel (2017), AMRUT has significantly contributed to improving water supply and sewerage systems in selected cities, which is crucial for sustainable urban development. Their analysis shows that cities covered under AMRUT have seen a noticeable improvement in basic service delivery, which is fundamental for enhancing urban living conditions.

AMRUT's emphasis on developing green spaces and promoting non-motorized transport is pivotal for environmental sustainability. A report by the National Institute of Urban Affairs (NIUA) (2018) underscores the positive outcomes of AMRUT's green initiatives, noting an increase in urban green cover and a reduction in urban heat islands. The initiative's focus on sustainable urban mobility, including the development of pedestrian-friendly pathways and cycling tracks, has also been well-received (Singh and Sharma, 2019). Effective urban planning and governance are critical for the success of AMRUT. According to a study by Kundu (2015), AMRUT has played a significant role in building the capacity of urban local bodies (ULBs) by providing training and resources necessary for effective urban management. The study highlights that the enhanced technical and managerial capabilities of ULBs have led to more efficient and sustainable urban planning practices.

Despite its successes, AMRUT has faced several challenges. Joshi and Saxena (2020) discuss the implementation challenges, including delays in project execution and issues related to inter-agency coordination. They argue that while AMRUT has laid a strong foundation for urban transformation, continuous monitoring, and adaptive management are required to overcome these hurdles. Comparative studies have also been conducted to assess AMRUT's performance relative to other urban development initiatives. Sharma and Sinha (2019) compare AMRUT with the Jawaharlal Nehru National Urban Renewal Mission (JNNURM), concluding that AMRUT's focused approach on basic infrastructure and its emphasis on sustainable practices provide a more robust framework for long-term urban sustainability.

Chakrabarti (2016) explores the policy implementation processes of AMRUT and their impact on urban reforms. The study highlights the administrative and procedural changes initiated under AMRUT, which aim to streamline project execution and improve governance at the municipal level. Mohan and Shashidhar (2018) examine the financial aspects and funding mechanisms of the AMRUT initiative. Their research investigates the allocation of resources, funding models, and financial sustainability of projects under AMRUT, offering critical insights into the economic viability of urban infrastructure projects.

Bhattacharya (2017) emphasizes the importance of public participation and community engagement in the AMRUT mission. The study discusses various strategies used to involve local communities in the planning and implementation processes, thereby enhancing the responsiveness and effectiveness of urban projects. Prasad (2019) investigates the role of AMRUT in enhancing urban resilience and disaster management. The study evaluates how AMRUT's infrastructure projects contribute to mitigating risks associated with natural disasters and climate change, promoting resilient urban development. Verma and Ramesh (2020) conduct a comparative analysis of AMRUT with similar urban policies in other countries. The study assesses the effectiveness of AMRUT in comparison to international best practices, providing a global perspective on urban sustainability initiatives.

Sharma and Kapoor (2018) assess the impact of AMRUT on urban health outcomes. Their research focuses on how improved water supply, sanitation, and waste management under AMRUT have contributed to better public health indicators in urban areas. Rao and Menon (2017) examine the gender and social inclusion aspects of AMRUT. This study looks at how the initiative addresses the needs of marginalized groups, including women, in urban planning and infrastructure development, ensuring equitable access to services. Kumar (2019) analyzes the integration of smart technologies in AMRUT projects. The study highlights the use of GIS, IoT, and other technological tools in urban planning and management, showcasing the move towards smart, data-driven urban governance. Jain and Singh (2021) provide a longitudinal study on the outcomes of AMRUT, tracking progress over time and projecting future impacts. Their research offers valuable insights into the long-term sustainability and scalability of urban development projects under AMRUT.

Tripathi and Jain (2017) investigate the improvements in urban water management under the AMRUT mission. Their study highlights the strategies adopted for efficient water distribution and wastewater treatment, emphasizing the environmental benefits and sustainability of these efforts. Desai and Shah (2019) focus on the advancements in solid waste management brought about by AMRUT. They assess the effectiveness of waste segregation, recycling, and disposal practices in selected cities, noting significant improvements in urban cleanliness and public health. Patel and Chatterjee (2020) analyze the correlation between infrastructure development under AMRUT and economic growth. Their research shows how enhanced urban infrastructure contributes to local economies by attracting investments and creating job opportunities. Khan and Ahmed (2018) provide detailed case studies of cities like Pune, Ahmedabad, and Lucknow to examine the localized impacts of AMRUT projects. These case studies offer insights into the unique challenges and successes experienced by different urban areas. Mehta and Roy (2019) explore how AMRUT initiatives contribute to climate change adaptation in urban environments. The study evaluates the resilience of urban infrastructure to climate-related risks, such as flooding and heat waves. Reddy and Kumar (2021) examine the integration of AMRUT with India's Smart Cities Mission. Their research highlights how the two initiatives complement each other in creating technologically advanced, sustainable urban areas.

Mishra and Das (2020) conduct a social impact assessment of AMRUT projects. The study focuses on the social benefits, such as improved living conditions and enhanced social equity, brought about by urban infrastructure improvements. Gupta and Bose (2019) provide a critical policy analysis of AMRUT. Their work includes recommendations for enhancing the effectiveness of urban planning and governance frameworks to ensure sustainable urban development. Chaudhary and Sharma (2018) explore the role of public-private partnerships (PPPs) in implementing AMRUT projects. Their study discusses the benefits and challenges of PPPs in urban infrastructure development. Nanda and Rao (2021) focus on improvements in urban mobility and transportation under AMRUT. They assess the development of public transport systems and non-motorized transport infrastructure, highlighting their contributions to reducing urban traffic congestion and pollution.

Research Gap and Objective of the Study

Thus, the literature review highlights a notable research gap regarding the identification of key factors contributing to sustainable urban planning and development, particularly within the AMRUT framework. Past studies have failed to adequately explain which factors are most crucial for fostering urban sustainability under the AMRUT program. This gap underscores the necessity for further investigation to understand the primary determinants influencing sustainable urban development under AMRUT. To achieve this goal, the research focuses on conducting a comprehensive investigation into the perceptions and evaluations of residents regarding urban environments, with particular emphasis on three cities in Odisha: Bhubaneswar, Cuttack, and Rourkela. The study is structured to address two central questions:

i. What are the parameters that help residents assess specific cities?
ii. What different weights can be assigned to these parameters based on resident responses?

To address these research questions, the study has developed three primary objectives:

- Identifying a conceptual model to understand the importance of various criteria for residents when deciding on a city,
- Assigning weights to these criteria scientifically based on empirical evidence gathered from residents.

Research Design

Data and Sample Selection

This study adopts a quantitative approach falling under the classification of multi-criteria decision-making methods (MCDM), specifically integrating the analytic hierarchy process (AHP) analysis. The research employs a stratified random sampling method to select the sample population. Targeting residents in cities governed by the AMRUT, the study encompasses 350 respondents from Bhubaneswar, Cuttack, and Rourkela over three months from October 2023 to December 2023. The key stakeholders in this research include residents of Bhubaneswar, Cuttack, and Rourkela, local authorities and policymakers associated with AMRUT, and the researchers conducting the study. Based on the AHP model a structured questionnaire was developed that serves as the primary instrument for data collection (Annexure 3). This questionnaire encompasses five distinct dimensions, comprising 32 factors as depicted in Figure 12.1 below (for details refer to Annexure 1). The respondents were tasked with comparing factors within each dimension against alternative factors. The fundamental AHP 9-point scale was applied for the pairwise comparisons (Table 12.1).

The Evaluation Model

For this research, the conceptual model of Ritchie and Crouch (2006) is being considered as a benchmark. The first reason for selecting this model is its comprehensiveness. The second reason for selecting this model is that it is the first and most cited model in destination competitiveness literature (Ayikoru, 2015). Third, this model has been widely adopted and empirically validated by researchers (Crouch, 2011). For the research, we are considering the 4 dimensions of the model proposed by Ritchie and Crouch (2006). We aim to measure the sustainability and effectiveness of the AMRUT cities from the citizen's point of view. The fifth dimension of this model is about the policymakers which does not involve the citizens directly and is hence excluded from the study that is based on the responses from the citizens. To facilitate the comparative evaluation of the cities of Bhubaneswar, Cuttack, and Rourkela, we employ the AHP. AHP, a MCDM tool, structures decision-making factors hierarchically and assigns weights to facilitate informed comparisons and evaluations.

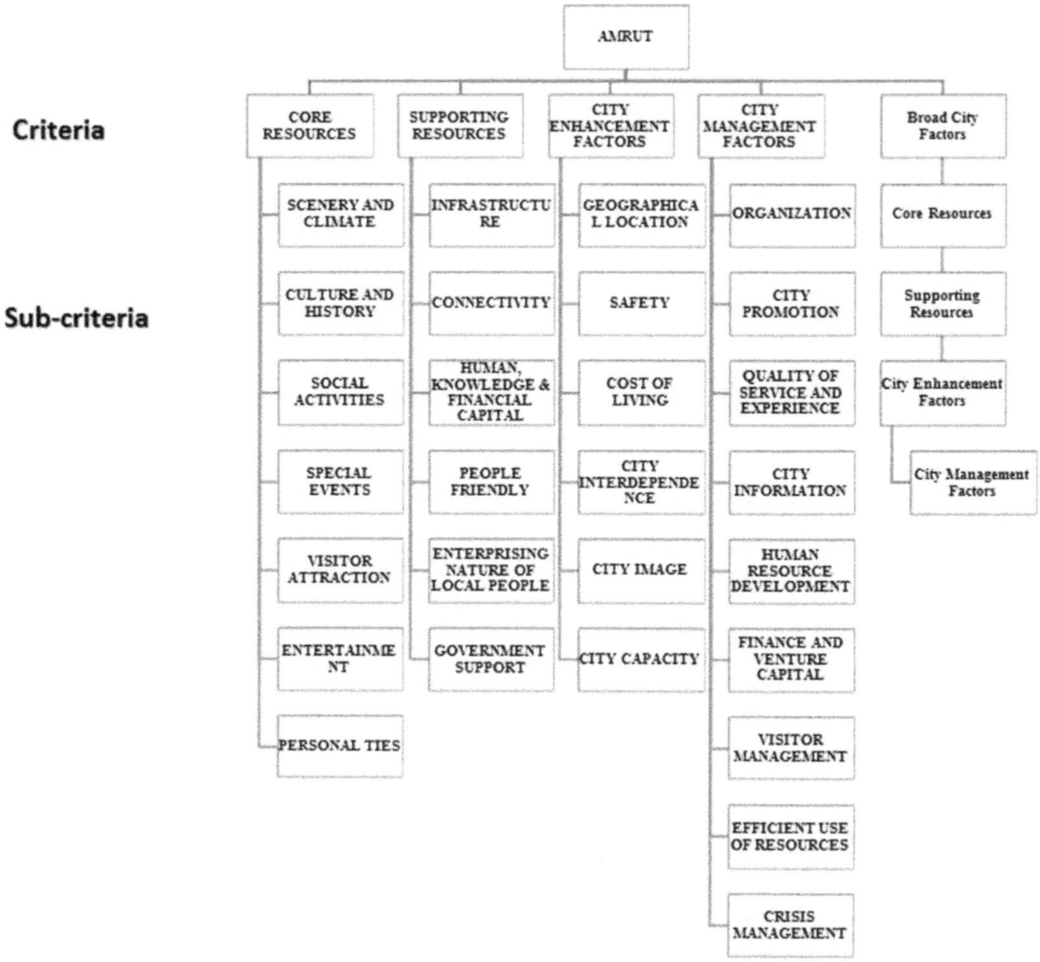

Figure 12.1 Analytical hierarchical model for AMRUT
Source: Computed by the authors

Table 12.1 AHP comparison scale.

Intensity of importance	Definition
1	Equal importance
3	Slightly importance
5	Strongly importance
7	Very strongly importance
9	Extremely importance
2,4,6,8	Intermediate values between adjacent judgments
Reciprocals of the above	If factor i has one of the above numbers assigned to it when compared to factor j, then j has the reciprocal value when compared with i

Source: Crouch, (2011), AHP scale importance

Data Analysis and Interpretation

Demographic Analysis

The demographic profile of the survey respondents for the AMRUT reveals several key characteristics. The majority of respondents are below 30 years of age, accounting for 75.94%, indicating a predominantly younger population. Gender distribution is relatively balanced, with a slight female majority (51.25%). Most respondents are unmarried (73.75%) and are spread across three cities: Bhubaneswar (37.50%), Cuttack (31.25%), and Rourkela (31.25%). The survey predominantly captures individuals with higher incomes, as 88.75% report earning above 5 lakhs annually. Educationally, 59.38% of respondents are graduates, while 40.63% have attained post-graduate or higher degrees. In terms of housing, a significant portion of respondents (65.63%) own their homes. Occupation-wise, the sample includes a large number of students (59.38%) and salaried employees (40.63%). Most respondents have substantial work experience, with 72.19% having more than five years in the workforce, and a similar percentage have lived in their respective cities for more than five years. This profile suggests that the survey captures the views of a young, well-educated, and economically stable urban population with significant work experience and long-term residency in their cities (Annexure 2).

AHP Model Analysis

The AHP model results highlight the significance of four key parameters: criteria weights, lambda max, consistency index (CI), and consistency ratio (CR). Criteria weights are pivotal in assessing and ranking alternatives based on their performance against established criteria. Lambda Max, ideally close to the number of criteria being compared, signifies strong consensus among decision-makers, with values exceeding the number of criteria by 10% considered acceptable. A lower CI (< 0.1) reflects minimal inconsistency in pairwise comparisons, indicating reliable judgments. Meanwhile, the CR compares the CI to a random benchmark, with values below 0.1 implying acceptable consistency. These parameters collectively provide insights into the robustness and reliability of decision-making processes within the AHP model. Each criterion of the model is interpreted in the following paragraphs.

Core Resources

Table 12.2 presents a pairwise comparison matrix and criteria weights for evaluating core resources in the context of city assessment. Scenery and climate, with a weight of 0.10, are moderately important in the AHP model, implying a role in urban planning but not as the primary focus within AMRUT. Culture and history, with a weight of 0.03, holds relatively low importance, suggesting historical and cultural aspects are not top priorities in urban development under AMRUT. Social activities, weighted at 0.21, emphasize community engagement and social interactions as significant factors. Special events, with a weight of 0.13, indicate consideration for occasional events but not as critical as other criteria. Visitor attraction, with a weight of 0.27, highlights a significant emphasis on aspects appealing to visitors, aligning with AMRUT's focus on tourism. Entertainment, at 0.09, suggests recognition but not primary focus, while personal ties, with a weight of 0.17, acknowledges factors contributing to social cohesion in urban planning within AMRUT.

In the AHP methodology applied to evaluate CR within the AMRUT framework, lambda max (λ_max) is calculated as 7.66, indicating a significant preference for certain core resources over others. This suggests a clear hierarchy or dominance in decision-making, underscoring the importance of these resources in urban development initiatives. The CI of 0.09 reflects a reasonable level of agreement among

Table 12.2 Criteria weights for core resources.

	Scenery and climate	Culture history	Social activities	Special events	Visitor attraction	Entertainment	Personal ties	Criteria weights
Scenery and climate	0.04	0.03	0.01	0.01	0.02	0.27	0.35	0.10
Culture history	0.04	0.03	0.01	0.01	0.02	0.05	0.02	0.03
Social activities	0.27	0.21	0.10	0.09	0.78	0.02	0.02	0.21
Special events	0.35	0.27	0.10	0.09	0.02	0.05	0.04	0.13
Visitor attraction	0.27	0.21	0.01	0.45	0.11	0.49	0.35	0.27
Entertainment	0.01	0.03	0.29	0.09	0.01	0.05	0.12	0.09
Personal ties	0.01	0.21	0.48	0.27	0.04	0.05	0.12	0.17

Source: Computed by the authors

Table 12.3 Criteria weights-supporting resources.

	Infrastructure	Connectivity	Human, knowledge and financial capital	People friendly	Enterprising nature of local people	Government support	Criteria weights
Infrastructure	0.03	0.02	0.01	0.02	0.03	0.04	0.02
Connectivity	0.23	0.11	0.33	0.02	0.67	0.19	0.26
Human, knowledge andfnancial capital	0.17	0.02	0.05	0.02	0.02	0.19	0.08
People friendly	0.23	0.74	0.24	0.11	0.02	0.19	0.25
Enterprising nature of local people	0.17	0.02	0.33	0.74	0.13	0.19	0.26
Government support	*0.17*	*0.11*	*0.05*	*0.11*	*0.13*	*0.19*	*0.13*

Source: Computed by the authors

stakeholders regarding the importance of core resources, contributing to informed decision-making processes. The CR of 0.07, falling below the threshold of 0.1, signifies an acceptable level of consistency in prioritizing core resources. This reaffirms the reliability of the decision-making process, emphasizing the foundational role of core resources in urban development within the AMRUT framework.

Supporting resources

Table 12.3 provides a pairwise comparison matrix of criteria weights for supporting

Table 12.4 Criteria weights-city enhancement factors.

	Geographical location	Safety	Cost of living	City interdependence	City Image	City capacity	Criteria weights
Geographical location	0.04	0.07	0.02	0.05	0.06	0.01	0.04
Safety	0.29	0.50	0.77	0.15	0.41	0.25	0.39
Cost of living	0.38	0.10	0.15	0.45	0.41	0.44	0.32
City Interdependence	0.04	0.17	0.02	0.05	0.06	0.01	0.06
City image	0.04	0.07	0.02	0.05	0.06	0.25	0.08
City capacity	0.21	0.10	0.02	0.25	0.01	0.05	0.11

Source: Computed by the authors

resources in the AHP model, crucial for urban development evaluation under AMRUT. Connectivity holds the highest weight of 0.26, emphasizing its pivotal role in advancing AMRUT goals through efficient urban networks. People friendly and enterprising nature of local people carry substantial weights of 0.25 and 0.26, respectively, aligning with AMRUT's focus on community-centric and entrepreneurial aspects. Government Support maintains a moderate weight of 0.13, indicating its consistent but relatively lesser role. Despite the lowest weight of 0.02, Infrastructure remains relevant, emphasizing its consideration in the AMRUT framework.

In the evaluation of supporting resources within the AMRUT framework, the Lambda Max (λ_max) value of 6.36 indicates a clear preference for the most dominant alternative, emphasizing the strategic importance of supporting resources. The low CI of 0.07 reflects a high level of agreement among decision-makers, ensuring reliable and coherent decision-making. Additionally, the CR of 0.06, well below the 0.1 threshold, confirms the robustness of the decision-making process. These metrics collectively highlight the critical role of supporting resources in enhancing foundational aspects and ensuring the sustainability of urban development initiatives (Table 12.7).

City Enhancement Factors

Table 12.4 presents a pairwise comparison matrix of criteria weights for city enhancement factors within the AHP model, crucial for evaluating cities under AMRUT. Safety holds the highest weight of 0.39, highlighting its critical role in ensuring secure urban spaces. Cost of living is also significant with a weight of 0.32, emphasizing the economic aspect of urban living. Geographical location, though lower at 0.04, remains relevant in decision-making. City interdependence has a weight of 0.06, reflecting collaborative aspects between cities. City Image carries a weight of 0.08, underscoring the importance of a positive city perception. City capacity, with a weight of 0.11, reflects the city's ability to accommodate growth sustainably.

In evaluating city enhancement factors within the AMRUT framework, the Lambda Max (λ_max) value of 6.54 indicates a clear preference for factors that improve urban amenities and aesthetics, highlighting their importance in enhancing the city's livability and attractiveness. The CI of 0.09 suggests a reasonable level of agreement among decision-makers, reflecting a coherent understanding of these factors' significance in improving quality of life. Additionally, the CR of 0.08, well below the 0.1 threshold, confirms the

Table 12.5 City management factors-criteria weights.

	Organization	City Promotion	Quality of service and experience	City information	Human resource development	Finance and venture capital	Visitor management	Efficient use of resources	Crisis management	Criteria weight
Organization	0.08	0.08	0.038	0.24	0.01	0.05	0.12	0.26	0.27	0.13
City promotion	0.08	0.08	0.113	0.02	0.21	0.27	0.04	0.03	0.16	0.11
Quality of service and experience	0.25	0.08	0.113	0.08	0.02	0.16	0.21	0.09	0.05	0.12
City information	0.03	0.39	0.113	0.08	0.35	0.02	0.12	0.03	0.05	0.13
Human resource development	0.41	0.03	0.338	0.02	0.07	0.16	0.12	0.03	0.05	0.14
Finance and venture capital	0.08	0.02	0.038	0.24	0.02	0.05	0.21	0.02	0.16	0.09
Visitor management	0.03	0.08	0.023	0.03	0.02	0.01	0.04	0.03	0.16	0.05
Efficient use of resources	0.03	0.23	0.113	0.24	0.21	0.27	0.12	0.09	0.01	0.15
Crisis management	0.02	0.03	0.11	0.08	0.07	0.02	0.01	0.43	0.05	0.09

Source: Computed by the authors

reliability and robustness of the decision-making process. These metrics collectively emphasize the strategic importance of city enhancement factors in urban development initiatives (Table 12.7).

City Management Factors

Table 12.5 provides an overview of criteria weights for city management factors under AMRUT, using the AHP model. The criteria cover essential aspects for effective city management. The weights indicate the relative importance of each criterion. Organization (0.13) highlights the efficiency of city management structures. City promotion (0.11) emphasizes activities attracting investments and tourism. Quality of service and experience (0.12) ensures a positive resident and visitor experience. City information (0.13) focuses on transparent information dissemination. human resource development (0.14) stresses investing in the city's workforce. Finance and venture capital (0.09) consider financial strategies for development. Visitor management (0.05) optimizes visitor experiences. efficient use of resources (0.15) prioritizes resource allocation for sustainability. Crisis Management (0.09) addresses planning for crisis and emergencies.

In the evaluation of city management factors using the AHP methodology, a Lambda Max (λ_max) value of 10.06 indicates a clear preference for governance, planning, and administrative factors, essential for effective urban management. CI of 0.09 reflects a reasonable level of agreement among decision-makers, highlighting the importance of coherent decision-making in urban governance. Furthermore, the CR of 0.08, well below the 0.1 threshold, confirms the acceptable consistency in prioritizing these factors. These metrics collectively underscore the critical role of city

Table 12.6 Broad city factors- criteria weights.

	Core resources	Supporting factors	City enhancement factors	City management factors	Criteria weights
Core resources	0.1	0.035715561	0.083333333	0.300030003	0.129769724
Supporting factors	0.3	0.107146684	0.083333333	0.100010001	0.147622505
City enhancement factors	0.5	0.535733419	0.416666667	0.300030003	0.438107522
City management factors	0.1	0.321440051	0.416666667	0.300030003	0.28453418

Source: Computed by the authors

Table 12.7 AHP model results for evaluation of AMRUT projects.

	Lambda max	Consistency index (CI)	Consistency ratio (CR)
Core resources	7.66	0.09	0.07
Supporting resources	6.36	0.07	0.06
City enhancement factors	6.54	0.09	0.08
City management factors	10.06	0.09	0.08
Broad city factors	4.98	0.03	0.02

Source: Computed by the authors

management factors in ensuring efficient governance and sustainable urban development (Table 12.7).

Broad City Factors

Table 12.6 presents criteria weights for different factors of city development within the AMRUT framework. Core resources carry a weight of 0.1298, emphasizing the importance of foundational infrastructure and services. Supporting Factors hold a weight of 0.1476, indicating their role in complementing core resources. City Enhancement Factors have the highest weight at 0.4381, highlighting their paramount importance in improving the overall quality of life and attractiveness of the city. City management factors are assigned a weight of 0.2845, underscoring the significance of effective governance and administration.

In the evaluation of broad city factors using the AHP methodology, a Lambda Max (λ_max) value of 4.98 indicates a clear preference for factors with wide-ranging implications for urban development. The CI)of 0.03 reflects a high level of consistency among decision-makers, highlighting the coherent understanding of the significance of these factors. Additionally, the CR of 0.02, well below the 0.1 threshold, confirms the high consistency in prioritizing broad city factors. These metrics collectively emphasize the critical role of these factors in decision-making processes aimed at fostering comprehensive and sustainable urban growth.

Discussion

The findings from the AHP analysis provide valuable insights into the prioritization of factors influencing urban development within the AMRUT framework. The significant Lambda Max (λ_max) values across various criteria underscore the clear dominance and hierarchy observed in decision-making, reflecting the crucial role of certain factors in shaping urban development initiatives under AMRUT. For instance, the high Lambda Max values for core resources indicate a strong preference for foundational elements such as infrastructure and basic services, highlighting their indispensable role in urban development planning. Moreover, the CI values demonstrate a reasonable level of agreement among stakeholders regarding the importance of these core resources, suggesting a shared understanding of their significance. This consensus is essential for effective decision-making and ensuring alignment with the objectives of the AMRUT initiative. The CR values falling below the threshold of 0.1 further validate the consistency and reliability of the decision-making process, affirming the robustness of the prioritization of core resources.

In contrast, factors related to city enhancement and management exhibit varying degrees of importance, as indicated by their respective Lambda Max values. While city enhancement factors carry the highest weights, emphasizing the importance of initiatives aimed at enhancing the overall quality of life and attractiveness of the city, city management factors also play a crucial role in ensuring efficient governance and administration. The consistency observed in prioritizing these factors underscores their collective significance in driving sustainable urban development within the AMRUT context. Overall, the study highlights the strategic focus on foundational elements, the importance of community-centric and entrepreneurial aspects, and the critical role of efficient governance and administration in urban development initiatives under AMRUT. These findings provide valuable insights for policymakers, urban planners, and stakeholders involved in shaping the future path of cities under the AMRUT initiative, facilitating informed decision-making and targeted interventions to address key challenges and opportunities in urban development.

Conclusion

In conclusion, our study provides a claer understanding of the Atal Mission for Rejuvenation and Urban Transformation (AMRUT), recognizing its pivotal role in addressing immediate societal needs while concurrently laying the groundwork for sustainable urban development. Through an exploration of residents' perspectives and an analysis of urban development factors, our research contributes valuable insights aimed at enhancing both the quality of life and economic prosperity within urban centers. We underscore the significance of prioritizing core resources, such as infrastructure and basic services, as foundational elements for sustainable urban growth, as highlighted by the high Lambda Max values and reasonable consistency index (CI) values. Moreover, the prominence of city enhancement factors emphasizes the importance of initiatives geared towards improving overall urban livability. Concurrently, the recognition of city management factors underscores the critical role of effective governance and administration in driving urban development initiatives forward. Looking ahead, future research studies should explore the long-term impact of AMRUT initiatives, conducting comparative analyses across states to identify best practices, explore the integration of emerging technologies, undertaking socio-economic studies, and assessing urban infrastructure resilience to climate change. By embracing these avenues of inquiry, policymakers and urban planners can effectively direct urban development toward a more sustainable, inclusive, and resilient future.

Acknowledgment

This article is an outcome of the short term project sponsored by Indian Council of Social Science Research (ICSSR) New Delhi under Special Call for Short-term Empirical Research 2023-24 (ICSSR-CIS-2023-1626). The authors thankful to ICSSR New Delhi for the support.

References

Ayikoru, M. (2015). Destination competitiveness challenges: A Ugandan perspective. Tourism Management, 50, 142–158.

Bhagat, R. B. (2012). A turnaround in India's urbanization. *Asia-Pacific Population Journal*, 27(2), 23–39.

Bhandari, L. (2016). AMRUT: a step towards sustainable urbanization in India. *Economic and Political Weekly*, 51(3), 23–27.

Bhattacharya, T. (2017). Public participation in urban planning: a case study of AMRUT. *Journal of Urban Development*, 12(4), 290–305.

Chakrabarti, S. (2016). Policy implementation and urban reforms under AMRUT. *Policy and Society*, 35(3), 287–302.

Chaudhary, S., and Sharma, K. (2018). Public-private partnerships in AMRUT: opportunities and challenges. *Journal of Infrastructure Development*, 10(3), 215–230.

Confederation of Indian Industry, Since 1895.

Crouch, G. I. (2011). A model of destination competitiveness and sustainability, 12 (1), 326–354.

Desai, M., and Shah, A. (2019). Advancements in solid waste management under AMRUT. *Environmental Management Journal*, 27(2), 142–158.

Gupta, A., and Bose, P. (2019). Policy analysis of AMRUT: enhancing urban planning and governance. *Urban Policy and Research*, 37(4), 320–336.

Jain, A., and Singh, N. (2021). Longitudinal study of AMRUT outcomes: tracking progress and future projections. *Urban Policy Review*, 29(1), 33–49.

Joshi, R., and Saxena, A. (2020). Implementation challenges in the AMRUT mission. *Journal of Urban Affairs*, 42(5), 707–722.

Khan, M., and Ahmed, S. (2018). Case studies of AMRUT cities: Pune, Ahmedabad, and Lucknow. *Urban Studies*, 55(6), 1352–1370.

Kumar, A. (2019). Technological integration in AMRUT: moving towards smart cities. *Journal of Urban Technology*, 26(2), 151–168.

Kundu, D. (2015). Urbanization trends of Indian metropolises: a case of Delhi with specific reference to the urban poor. *Urban Development Challenges, Risks and Resilience in Asian Mega Cities*, 31–47.

Mehta and Roy, (2019). "Role of AMRUT Project in Urban Governance of West Bengal: A Review of Barddhaman Municipality (2015-2020)." Khazanah Sosial 5, no. 1: 37–52.

Mishra, R., and Das, S. (2020). Social impact assessment of AMRUT projects. *Social Development Review*, 34(1), 76–92.

Mohan, M., and Kandya, A. (2015). Impact of urbanization and land-use/land-cover change on diurnal temperature range: a case study of tropical urban airshed of India using remote sensing data. *Science of the Total Environment*, 506, 453–465.

Mohan, R., and Shashidhar, S. (2018). Financial aspects and funding mechanisms of AMRUT. *Finance and Urban Development Journal*, 18(2), 76–89.

Nanda, P., and Rao, V. (2021). Urban mobility improvements under AMRUT. *Transportation Research Journal*, 16(3), 203–219.

National Institute of Urban Affairs (NIUA) (2018). Green Spaces and AMRUT: Enhancing Urban Environments. NIUA Publications.

Patel, D., and Chatterjee, S. (2020). Infrastructure development and economic growth under AMRUT. *Economic Development Quarterly*, 34(2), 129–145.

Prasad, N. (2019). Urban resilience and disaster management in AMRUT Cities. *International Journal of Disaster Risk Reduction*, 31, 235–243.

Rao, P. V., and Patel, S. (2017). Infrastructure development under AMRUT: a review. *Journal of Infrastructure Development*, 9(1), 42–58.

Rao, S., and Menon, S. (2017). Gender and social inclusion in AMRUT projects. *Journal of Social Inclusion Studies*, 23(1), 112–126.

Reddy, B., and Kumar, P. (2021). Integration of AMRUT with smart cities mission. *International Journal of Smart City Research*, 9(1), 43–59.

Ritchie, J. R. B., and Crouch, G. I. (2003). The Competitive Destination: A Sustainable Tourism Perspective. CABI.

Sharma, A., and Sinha, R. (2019). Comparing AMRUT and JNNURM: a study of urban policy frameworks. *Urban Studies Review*, 56(3), 378–395.

Sharma, V., and Kapoor, S. (2018). Impact of AMRUT on urban health outcomes. *Urban Health Journal*, 14(3), 204–219.

Singh, R., and Sharma, P. (2019). Sustainable urban mobility under AMRUT: an evaluation. *International Journal of Sustainable Development and Planning*, 14(2), 189–200.

Tripathi, S., and Jain, R. (2017). Urban water management improvements under AMRUT. *Water Resources Management Journal*, 31(5), 1325–1340.

Verma, G., and Ramesh, A. (2020). Comparative analysis of AMRUT and international urban policies. *Global Urban Studies Review*, 11(4), 412–429.

Annexures

Annexure 12.1 Factors used in the AHP survey for evaluation of AMRUT projects.

Factors	Definition
Core resources	
Scenery and Climate	With its captivating scenery and pleasant climate, the city offers an enticing place to live.
Culture and history	The city's rich cultural heritage and history create an inviting environment to make it home.
Social activities	The city's dynamic events and vibrant festivals foster creativity, unity, and a strong sense of community identity.
Special events	A city's special events showcase its diverse heritage, from world fairs to cultural festivals.
Visitor attraction	This city's historic landmarks, museums, parks, diverse cuisine, etc., make it a memorable destination for visitors.
Entertainment	The city's arts and entertainment scene, with theaters, music venues, and recreational options, fosters creativity and community engagement.
Personal ties	The city's people share strong family and friendship bonds, united by common religious and ethnic roots.
Supporting resources	
Infrastructure	The city possesses a strong infrastructure, including reliable transportation, safety, utilities, healthcare, education, and retail services, meeting the diverse needs of its residents effectively.
Connectivity	The city offers accessible transportation options and visitor-friendly security measures for a hassle-free experience.
Human, knowledge and financial capital	The city's education and research institutions cultivate skilled professionals, while its financial & banking services support economic growth and stability.
People friendly	The city's residents warmly welcome outsiders, reflecting genuine hospitality and cultural values.
Enterprising nature of local people	The city's people are highly entrepreneurial, embracing risk, innovation, and adaptability, fostering a dynamic and forward-thinking business environment.
Government support	Government support drives infrastructure, economy, and well-being, covering public services, education, healthcare, and sustainability, ensuring the city's progress.
City enhancement factors	
Geographical location	The city's ideal geographical location offers beauty, urban convenience, and a comfortable year-round climate, making it a perfect home.
Safety	The city ensures safety through law enforcement, urban planning, and community engagement, creating a secure environment for all.
Cost of living	The city provides an attractive, balanced cost of living with affordable housing, accessible healthcare, diverse entertainment, ample education, and job opportunities for a comfortable lifestyle.
City interdependence	Cities maintain interdependent relationships, to shared strengths for mutual benefit and regional growth, like twin cities.
City image	The perception, awareness, and image of a city in the minds of people.

Factors	Definition
City capacity	The city's infrastructure limits its ability to meet growing demands, emphasizing the need for sustainable development and future planning.
City management factors	
Organization	The city's organization fosters efficient governance and service delivery, promoting coordination across various departments to cater to the needs of its residents.
City promotion	City promotion involves strategically showcasing the city's unique attributes and cultural richness to a global audience through various marketing initiatives and cultural events, fostering economic growth and global recognition.
Quality of service and experience	The city provides top-notch services and amenities, creating a welcoming atmosphere for residents and visitors.
City information	The residents/locals of a city know about the city services, and locations and are well aware of the available amenities.
Human resource development	Human resource development in the city involves nurturing a skilled workforce through diverse education and training initiatives.
Finance and venture capital	The city's finance and venture capital sectors foster economic growth through diverse financial services and robust support for startups.
Visitor management	City visitor management ensures efficient entry through registration, security, and technology, aiming for a secure and organized urban experience. It involves crowd control, communication, and continuous improvement for a better visitor experience.
Efficient use of resources	The city emphasizes efficient resource usage through innovative practices, that include recycling programs, sustainable infrastructure, and eco-friendly policies.
Crisis management	The city's crisis management employs proactive strategies, robust communication, and swift resource deployment to ensure the safety and well-being of its residents during challenging times such as cyclones, floods, etc.
Broad city factors	
Core resources	The city's charm lies in captivating scenery and a pleasant climate, offering an inviting atmosphere rich in cultural heritage and history. Dynamic social activities, vibrant festivals, and diverse special events showcase the city's identity, making it memorable for both residents and visitors. The arts and entertainment scene fosters creativity, while strong family and friendship bonds, rooted in common religious and ethnic ties, emphasize social cohesion within the community.
Supporting resources	The city's strength lies in robust infrastructure meeting diverse resident needs, with reliable transportation, safety measures, utilities, healthcare, education, and retail services. Offering hassle-free experiences, it provides accessible transportation and visitor-friendly security. Fuelled by developed human, knowledge, and financial capital, education and research institutions cultivate skilled professionals, while financial services support economic growth. Residents' welcoming nature, combined with a highly entrepreneurial population, fosters a dynamic and innovative business environment. Government support drives infrastructure development, economic growth, and overall well-being through initiatives in public services, education, healthcare, and sustainability, ensuring continuous progress.

Factors	Definition
City enhancement factors	The city's appeal is rooted in an ideal location, offering beauty, urban convenience, and a comfortable year-round climate. Safety is ensured through effective law enforcement and urban planning. A balanced cost of living, with affordable housing and accessible amenities, contributes to a comfortable lifestyle. Interdependence with twin cities fosters shared strengths for mutual benefit, while managing city capacity requires sustainable development and future planning for continued progress. The perception and image of the city shape its identity.
City management factors	Effective city management ensures seamless governance and service delivery, meeting diverse resident needs. City promotion strategically showcases unique attributes, fostering economic growth and global recognition. Top-notch services create a welcoming atmosphere for residents and visitors. Well-informed locals contribute to the city's success, and human resource development prioritizes a skilled workforce. The finance sector supports economic growth, while visitor management ensures a smooth experience. The city emphasizes efficient resource usage through sustainable practices and proactive crisis management ensures resident safety during challenging times.

Source: Compiled by authors

Annexure 12.2 Demographic details.

Variables	Categories	Sample	Proportion (%)
Age	Below 30	243	75.94
	Above 30	77	24.06
Gender	Male	156	48.75
	Female	164	51.25
Marital status	Married	82	25.63
	Unmarried	236	73.75
Location	Bhubaneswar	120	37.50
	Cuttack	100	31.25
	Rourkela	100	31.25
Income	Below 5 lakhs	36	11.25
	Above 5 lakhs	284	88.75
Education	Graduate	190	59.38
	Post Graduation & above	130	40.63
Housing status	Own house	210	65.63
	Rented House	110	34.38
Occupation	Student	190	59.38
	Salaried Employee	130	40.63
Work experience	Below 5 years	89	27.81
	Above 5 years	231	72.19
Duration in the city	Below 5 years	89	27.81
	Above 5 years	231	72.19

Source: Compiled by authors

Annexure 12.3 Questionnaire.

A- Core Resources.

	Extremely Important		Very strongly important		Strongly important		Slightly important		Equal		Slightly important		Strongly important		Very strongly important		Extremely important	
Scenery & climate	9	8	7	6	5	4	3	2	1	2	3	4	5	6	7	8	9	Culture & history
Scenery & climate	9	8	7	6	5	4	3	2	1	2	3	4	5	6	7	8	9	Social activities
Scenery & climate	9	8	7	6	5	4	3	2	1	2	3	4	5	6	7	8	9	Special events
Scenery & climate	9	8	7	6	5	4	3	2	1	2	3	4	5	6	7	8	9	Visitor attraction
Scenery & climate	9	8	7	6	5	4	3	2	1	2	3	4	5	6	7	8	9	Entertainment
Scenery & climate	9	8	7	6	5	4	3	2	1	2	3	4	5	6	7	8	9	Personal ties
Culture & history	9	8	7	6	5	4	3	2	1	2	3	4	5	6	7	8	9	Social activities
Culture & history	9	8	7	6	5	4	3	2	1	2	3	4	5	6	7	8	9	Special events
Culture & history	9	8	7	6	5	4	3	2	1	2	3	4	5	6	7	8	9	Visitor attraction
Culture & history	9	8	7	6	5	4	3	2	1	2	3	4	5	6	7	8	9	Entertainment
Culture & history	9	8	7	6	5	4	3	2	1	2	3	4	5	6	7	8	9	Personal ties
Social activities	9	8	7	6	5	4	3	2	1	2	3	4	5	6	7	8	9	Special events
Social activities	9	8	7	6	5	4	3	2	1	2	3	4	5	6	7	8	9	Visitor attraction
Social activities	9	8	7	6	5	4	3	2	1	2	3	4	5	6	7	8	9	Entertainment
Social activities	9	8	7	6	5	4	3	2	1	2	3	4	5	6	7	8	9	Personal ties
Special events	9	8	7	6	5	4	3	2	1	2	3	4	5	6	7	8	9	Visitor attraction
Special events	9	8	7	6	5	4	3	2	1	2	3	4	5	6	7	8	9	Entertainment
Special events	9	8	7	6	5	4	3	2	1	2	3	4	5	6	7	8	9	Personal ties
Visitor attraction	9	8	7	6	5	4	3	2	1	2	3	4	5	6	7	8	9	Entertainment
Visitor attraction	9	8	7	6	5	4	3	2	1	2	3	4	5	6	7	8	9	Personal ties
Entertainment	9	8	7	6	5	4	3	2	1	2	3	4	5	6	7	8	9	Personal ties

Source: Compiled by authors

B- Supporting Resources.

	Extremely important		Very strongly important		Strongly important		Slightly important		Equal		Slightly important		Strongly important		Very strongly important		Extremely important	
Infrastructure	9	8	7	6	5	4	3	2	1	2	3	4	5	6	7	8	9	Connectivity
Infrastructure	9	8	7	6	5	4	3	2	1	2	3	4	5	6	7	8	9	Human, knowledge, & financial capital
Infrastructure	9	8	7	6	5	4	3	2	1	2	3	4	5	6	7	8	9	People friendly
Infrastructure	9	8	7	6	5	4	3	2	1	2	3	4	5	6	7	8	9	Enterprising nature of local people
Infrastructure	9	8	7	6	5	4	3	2	1	2	3	4	5	6	7	8	9	Government support
Connectivity	9	8	7	6	5	4	3	2	1	2	3	4	5	6	7	8	9	Human, knowledge, & financial capital
Connectivity	9	8	7	6	5	4	3	2	1	2	3	4	5	6	7	8	9	People friendly
Connectivity	9	8	7	6	5	4	3	2	1	2	3	4	5	6	7	8	9	Enterprising nature of local people
Connectivity	9	8	7	6	5	4	3	2	1	2	3	4	5	6	7	8	9	Government support
Human, knowledge, & financial capital	9	8	7	6	5	4	3	2	1	2	3	4	5	6	7	8	9	People friendly
Human, knowledge, & financial capital	9	8	7	6	5	4	3	2	1	2	3	4	5	6	7	8	9	Enterprising nature of local people
Human, knowledge, & financial capital	9	8	7	6	5	4	3	2	1	2	3	4	5	6	7	8	9	Government support
People friendly	9	8	7	6	5	4	3	2	1	2	3	4	5	6	7	8	9	Enterprising nature of local people
People friendly	9	8	7	6	5	4	3	2	1	2	3	4	5	6	7	8	9	Government support
Enterprising nature of local people	9	8	7	6	5	4	3	2	1	2	3	4	5	6	7	8	9	Government support

Source: Compiled by authors

C- City Enhancement Factors.

	Extremely Important		Very strongly important		Strongly important		Slightly important		Equal		Slightly important		Strongly important		Very strongly important		Extremely important	
Geographical location	9	8	7	6	5	4	3	2	1	2	3	4	5	6	7	8	9	Safety
Geographical location	9	8	7	6	5	4	3	2	1	2	3	4	5	6	7	8	9	Cost of Living
Geographical location	9	8	7	6	5	4	3	2	1	2	3	4	5	6	7	8	9	City Interdependence
Geographical location	9	8	7	6	5	4	3	2	1	2	3	4	5	6	7	8	9	City Image
Geographical location	9	8	7	6	5	4	3	2	1	2	3	4	5	6	7	8	9	City capacity
Safety	9	8	7	6	5	4	3	2	1	2	3	4	5	6	7	8	9	Cost of Living
Safety	9	8	7	6	5	4	3	2	1	2	3	4	5	6	7	8	9	City Interdependence
Safety	9	8	7	6	5	4	3	2	1	2	3	4	5	6	7	8	9	City Image
Safety	9	8	7	6	5	4	3	2	1	2	3	4	5	6	7	8	9	City capacity
Cost of living	9	8	7	6	5	4	3	2	1	2	3	4	5	6	7	8	9	City Interdependence
Cost of living	9	8	7	6	5	4	3	2	1	2	3	4	5	6	7	8	9	City Image
Cost of living	9	8	7	6	5	4	3	2	1	2	3	4	5	6	7	8	9	City capacity
City interdependence	9	8	7	6	5	4	3	2	1	2	3	4	5	6	7	8	9	City Image
City interdependence	9	8	7	6	5	4	3	2	1	2	3	4	5	6	7	8	9	City capacity
City image	9	8	7	6	5	4	3	2	1	2	3	4	5	6	7	8	9	City capacity

Source: Compiled by authors

D- City Management Factors.

	Extremely Important		Very strongly important		Strongly important		Slightly important		Equal		Slightly important		Strongly important		Very strongly important		Extremely important	
Organization	9	8	7	6	5	4	3	2	1	2	3	4	5	6	7	8	9	City promotion
Organization	9	8	7	6	5	4	3	2	1	2	3	4	5	6	7	8	9	Quality of service and experience
Organization	9	8	7	6	5	4	3	2	1	2	3	4	5	6	7	8	9	City information
Organization	9	8	7	6	5	4	3	2	1	2	3	4	5	6	7	8	9	Human resource development
Organization	9	8	7	6	5	4	3	2	1	2	3	4	5	6	7	8	9	Finance and Venture Capital
Organization	9	8	7	6	5	4	3	2	1	2	3	4	5	6	7	8	9	Visitor management
Organization	9	8	7	6	5	4	3	2	1	2	3	4	5	6	7	8	9	Efficient use of resources
Organization	9	8	7	6	5	4	3	2	1	2	3	4	5	6	7	8	9	Crisis management
City Promotion	9	8	7	6	5	4	3	2	1	2	3	4	5	6	7	8	9	Quality of service and experience
City Promotion	9	8	7	6	5	4	3	2	1	2	3	4	5	6	7	8	9	City information
City Promotion	9	8	7	6	5	4	3	2	1	2	3	4	5	6	7	8	9	Human resource development
City Promotion	9	8	7	6	5	4	3	2	1	2	3	4	5	6	7	8	9	Finance and venture capital
City Promotion	9	8	7	6	5	4	3	2	1	2	3	4	5	6	7	8	9	Visitor management
City Promotion	9	8	7	6	5	4	3	2	1	2	3	4	5	6	7	8	9	Efficient use of resources
City Promotion	9	8	7	6	5	4	3	2	1	2	3	4	5	6	7	8	9	Crisis management
Quality of service & experience	9	8	7	6	5	4	3	2	1	2	3	4	5	6	7	8	9	City information

	Extremely Important		Very strongly important		Strongly important		Slightly important		Equal		Slightly important		Strongly important		Very strongly important		Extremely important	
Quality of service & experience	9	8	7	6	5	4	3	2	1	2	3	4	5	6	7	8	9	Human resource development
Quality of service & experience	9	8	7	6	5	4	3	2	1	2	3	4	5	6	7	8	9	Finance and venture capital
Quality of service & experience	9	8	7	6	5	4	3	2	1	2	3	4	5	6	7	8	9	Visitor management
Quality of service & experience	9	8	7	6	5	4	3	2	1	2	3	4	5	6	7	8	9	Efficient use of resources
Quality of service & experience	9	8	7	6	5	4	3	2	1	2	3	4	5	6	7	8	9	Crisis management
City information	9	8	7	6	5	4	3	2	1	2	3	4	5	6	7	8	9	Human resource development
City information	9	8	7	6	5	4	3	2	1	2	3	4	5	6	7	8	9	Finance and venture capital
City information	9	8	7	6	5	4	3	2	1	2	3	4	5	6	7	8	9	Visitor management
City information	9	8	7	6	5	4	3	2	1	2	3	4	5	6	7	8	9	Efficient use of resources
City information	9	8	7	6	5	4	3	2	1	2	3	4	5	6	7	8	9	Crisis management
Human resource development	9	8	7	6	5	4	3	2	1	2	3	4	5	6	7	8	9	Finance and venture capital
Human resource development	9	8	7	6	5	4	3	2	1	2	3	4	5	6	7	8	9	Visitor management
Human resource development	9	8	7	6	5	4	3	2	1	2	3	4	5	6	7	8	9	Efficient use of resources
Human resource revelopment	9	8	7	6	5	4	3	2	1	2	3	4	5	6	7	8	9	Crisis management
Finance and venture capital	9	8	7	6	5	4	3	2	1	2	3	4	5	6	7	8	9	Visitor management

	Extremely Important		Very strongly important		Strongly important		Slightly important		Equal		Slightly important		Strongly important		Very strongly important		Extremely important	
Finance and venture capital	9	8	7	6	5	4	3	2	1	2	3	4	5	6	7	8	9	Efficient use of resources
Finance and venture capital	9	8	7	6	5	4	3	2	1	2	3	4	5	6	7	8	9	Crisis management
Visitor management	9	8	7	6	5	4	3	2	1	2	3	4	5	6	7	8	9	Efficient use of resources
Visitor management	9	8	7	6	5	4	3	2	1	2	3	4	5	6	7	8	9	Crisis management
Efficient use of resources	9	8	7	6	5	4	3	2	1	2	3	4	5	6	7	8	9	Crisis management

Source: Compiled by authors

E- Broad City Factors.

	Extremely important		Very strongly important		Strongly important		Slightly important		Equal		Slightly important		Strongly important		Very strongly important		Extremely important	
Core resources	9	8	7	6	5	4	3	2	1	2	3	4	5	6	7	8	9	Supporting factors
Core resources	9	8	7	6	5	4	3	2	1	2	3	4	5	6	7	8	9	City enhancement factors
Core resources	9	8	7	6	5	4	3	2	1	2	3	4	5	6	7	8	9	City management factors
Supporting factors	9	8	7	6	5	4	3	2	1	2	3	4	5	6	7	8	9	City enhancement factors
Supporting factors	9	8	7	6	5	4	3	2	1	2	3	4	5	6	7	8	9	City management factors
City enhancement factors	9	8	7	6	5	4	3	2	1	2	3	4	5	6	7	8	9	City management factors

Source: Compiled by authors

13 Innovative business models and strategies for a sustainable future: an empirical analysis

Sneha Lata Das[a]

Assistant Professor, STRM(Commerce), KISS Deemed to be University, Bhubaneswar, India

Abstract

This empirical research paper undertakes a rigorous investigation into the intersection of innovative business models and sustainable strategies, aiming to provide a comprehensive understanding of their real-world applications and impacts within an increasingly environmentally conscious business landscape. Through an extensive literature review, a robust theoretical framework is constructed to guide the empirical analysis. Leveraging a mixed-methods approach, which encompasses surveys and interviews, authors investigate deeply into the adoption and performance of innovative business models across a spectrum of industries. Data analysis uncovers the relationships between sustainable strategies and business outcomes, shedding light on previously unexplored connections. These findings not only enrich the existing body of knowledge on sustainable business practices but also offer actionable insights for businesses looking to effectively integrate innovation and sustainability into their operations. By clarifying the dynamics between innovative business models and a sustainable future, this research facilitates informed decision-making and strategic planning in the business. The significance of this study lies in its contribution to a holistic understanding of the intricate interplay between innovation and sustainability, paving the way for more informed and conscientious business practices. By highlighting the practical implications of this intersection, author aim to empower businesses to embrace sustainable strategies while fostering innovation, thereby advancing the discourse on corporate responsibility and environmental stewardship in contemporary commerce. This research seeks to equip businesses with the knowledge and tools necessary to navigate the complexities of the modern business environment, promoting a more sustainable and resilient future for all stakeholders involved.

Keywords: Business performance, environmental consciousness, innovative business models, sustainable business practices, sustainable strategies

Introduction

In the unfolding narrative of our global trajectory, recent decades have borne witness to an unprecedented surge in challenges that traverse the realms of society and the environment. From the ominous specter of climate change, with its rising temperatures and cascading impacts, to the depletion of vital natural resources amid a burgeoning global population, the landscape is rife with interwoven crises. Moreover, social inequality, manifesting as widening gaps in income and resource access, and geopolitical instability further amplify the complexity of the issues at hand. In the face of these challenges, a clarion call resounds—a call for a fundamental shift in the modus operandi of businesses. The looming threat of climate change casts a long shadow over our world, bringing with it not only rising temperatures but also a symphony of extreme weather events and far-reaching impacts on ecosystems and human societies. Simultaneously, the depletion of natural resources, intensified by a burgeoning global population, has sparked

[a]snehadasbbsr@gmail.com

DOI: 10.1201/9781003564447-13

heightened competition for essential commodities, giving rise to profound concerns about the long-term sustainability of our way of life. The fabric of our societies is further strained by social inequality, marked by yawning chasms in income and access to resources, necessitating urgent attention to issues of equity and justice. Amidst this tapestry of challenges, geopolitical instability, characterized by political and economic uncertainties on the global stage, underscores the imperative for businesses to navigate intricate and evolving environments. In this crucible of challenges, businesses emerge as central players endowed with the potential to effect transformative change. Beyond conventional profit-centric paradigms, there is a growing acknowledgment of the profound responsibility borne by businesses to contribute positively to both society and the environment. The pivotal role of the business sector extends beyond mere recognition of its impact on these challenges; it lies in the proactive pursuit of solutions that align with sustainable development goals. Within this dynamic landscape, the transformative potential of sustainable business models takes center stage. Encouraging innovation in products, services, and operations, these models foster heightened efficiency and reduced environmental impact. Stakeholder engagement becomes a cornerstone, with businesses prioritizing the interests not only of shareholders but also of employees, communities, and the environment. Integral to risk mitigation, sustainability strategies enable businesses to anticipate and adapt to changes in regulatory environments, consumer preferences, and global conditions. Above all, the crux of the transformative potential lies in the ability of sustainable business models to promote long-term resilience, ensuring the continued viability of businesses in the face of evolving challenges. In essence, the significance of sustainable business models lies not only in their capacity to address pressing societal and environmental issues but also in positioning businesses as active contributors to positive change. These models secure the long-term viability of businesses in a rapidly changing world. The subsequent empirical analysis endeavors to delve into the practical implications and effectiveness of these models in achieving these lofty goals.

Literature Review

In response to global challenges such as climate change, resource depletion, and social inequality, there has been a paradigm shift in the business landscape towards sustainable practices. Innovative business models, characterized by their adaptability and forward-thinking nature, have emerged as a crucial avenue for organizations to integrate sustainability into their core strategies. Numerous studies have explored the diversity of innovative business models across industries. For instance, Chesbrough (2010) work on open innovation sheds light on how collaborative models can foster sustainability through shared knowledge. Similarly, Teece (2010) concept of dynamic capabilities emphasizes the role of adaptability and innovation in achieving a sustainable competitive advantage. Examining sustainability strategies, Elkington (1997) Triple bottom line framework remains seminal in evaluating organizational success based on economic, social, and environmental outcomes. Additionally, Porter and Kramer (2011), Das (2022), Das and Das (2020) concept of creating shared value advocates for strategies that align business success with societal progress. Despite the richness of conceptual literature, empirical research on the practical implementation and impact of innovative business models remains limited. Few studies systematically investigate the link between specific innovative strategies and tangible sustainability outcomes. This literature gap highlights the need for empirical studies that bridge the theoretical-practical divide. Recent studies have underscored the pivotal role of innovative business models in addressing sustainability challenges across various industries. For instance, Smith et al. (2023) conducted a comprehensive analysis of circular economy business models, highlighting their potential to minimize resource depletion and environmental impact while fostering economic growth. Furthermore, Johnson and Lee (2024) explored the emergence of platform-based business models in the renewable energy sector, showcasing how digital platforms enable greater efficiency and transparency in energy distribution, thus contributing to sustainability goals. In parallel, the discourse on sustainability initiatives has evolved to emphasize the importance of holistic approaches that integrate economic, social, and environmental considerations. Garcia et al. (2022) examined the concept of regenerative agriculture as a sustainable business model, demonstrating its potential to enhance soil health, biodiversity, and farmer livelihoods while mitigating climate change impacts. Similarly, Chen and Wang (2023) investigated the role of sustainable supply chain practices in the fashion industry, emphasizing the need for collaboration and transparency among stakeholders to achieve long-term sustainability objectives. By encompassing a wider range of contemporary literature, this review aims to provide a total understanding of the latest trends and insights shaping the intersection of innovative business models and sustainability initiatives. Through empirical research and case studies, we seek to elucidate practical strategies and opportunities for organizations to drive positive environmental and social outcomes while maintaining competitive advantage in an increasingly complex business landscape.

For this study some hypotheses are made for testing. The proposed hypotheses collectively explore the intricate relationships between innovation, sustainability, and organizational performance. Hypothesis 1 posits a positive correlation between the level of innovation in business models and the degree of sustainability integration, suggesting that more innovative firms are likely to adopt sustainable practices. Hypothesis 2 highlights the importance of organizational factors such as leadership commitment and organizational culture in influencing the adoption of sustainable business models, indicating that internal dynamics play a crucial role. Hypothesis 3 asserts that organizations adopting innovative, sustainability-focused business models achieve positive outcomes in environmental impact, social responsibility, and economic performance. Furthermore, Hypothesis 4 suggests that the relationship between sustainable business model adoption and organizational performance is moderated by factors such as organizational size, industry type, and geographical location, pointing to the contextual nature of these relationships. Finally, Hypothesis 5 emphasizes that effective stakeholder engagement enhances the success of sustainable business models, leading to improved organizational outcomes, thus underscoring the importance of external interactions in driving sustainability initiatives. Together, these hypotheses offer a comprehensive framework for understanding the multi-faceted nature of sustainability in business.

Despite the growing body of literature on innovative business models and sustainability strategies, a notable research gap exists in understanding the detailed mechanisms and contextual factors influencing the successful implementation of these models in diverse organizational settings. While existing studies provide insights into the conceptual frameworks and theoretical underpinnings of innovative business models and sustainability strategies, there is a lack of empirical research that delves into the practical challenges and outcomes associated with their adoption. Specifically, there is a need for empirical studies that explore the contextual factors that facilitate or hinder the adoption and implementation of innovative business models for sustainability across different industries and regions. This study will investigate the role of organizational culture, leadership styles, and internal capabilities in driving the successful integration of sustainability principles into business strategies through innovative models. It also examine the effectiveness of different types of innovative business models (e.g., collaborative, circular, platform-based) in addressing specific sustainability challenges, such as reducing carbon emissions, enhancing resource efficiency, or promoting social equity. This will assess the long-term sustainability outcomes and performance implications of organizations that have embraced innovative business models compared to those following traditional approaches. it will identify best practices and practical strategies for overcoming barriers and maximizing the benefits of innovative business models for sustainability, drawing on real-world case studies and empirical evidence. Addressing these research gaps can provide valuable insights for practitioners, policymakers seeking to advance the adoption of sustainable business practices and drive meaningful change a more sustainable future. Moreover, bridging the divide between theory and practice through empirical research can contribute to the development of evidence-based strategies that foster innovation and sustainability in organizations across diverse contexts.

The concept of a circular economy, as detailed by Geissdoerfer et al. (2017), advocates for a sustainable approach by enhancing resource efficiency, minimizing waste, and extending product lifecycles. Complementing this, Schiederig et al. (2012) propose a framework for sustainable innovation, which underscores the necessity of incorporating environmental and social considerations into the innovation process. Together, these models emphasize the interconnectedness of resource sustainability and innovation in achieving long-term ecological and societal benefits.

In crafting the theoretical framework for this study, the author integrates insights from a comprehensive literature review on innovative business models (IBM) and sustainability strategies. By applying the resource-based view (RBV) perspective (Barney, 1991), the study examines how an organization's unique resources and capabilities enhance its sustainability efforts. Additionally, institutional theory, as developed by DiMaggio and Powell (1983), is employed to understand how institutional pressures influence organizations' decisions to adopt and legitimize sustainable practices. This framework serves as a compass, guiding the investigation into the intricate relationship between the adoption of innovative business models and the consequent realization of sustainability outcomes within organizations. While the literature review highlights a rich array of theoretical concepts related to IBM and sustainability strategies, it also reveals a scarcity of empirical studies—a critical gap that this research aims to fill. By exploring practical applications and outcomes, this study seeks to contribute valuable insights to the ongoing discourse on sustainable business practices.

Research Questions and Objectives

The study is anchored by a set of research questions that explore various dimensions of adopting innovative business models focused on sustainability. Group-1 questions assess the extent to which businesses across different industries adopt these models and identify key drivers and barriers influencing their adoption. This assessment provides a foundational understanding of the current landscape and contextualizes subsequent inquiries. Group-2 investigates how these models impact organizations' environmental, social, and economic performance, which is critical for evaluating their effectiveness. Group-3 examines variations in adoption and outcomes based on organizational size, industry type, or geographical location, helping to tailor strategies and interventions to specific contexts. Finally, Group-4 investigates into the role of stakeholder engagement in the successful implementation of sustainable business models, emphasizing its importance in aligning with stakeholders' expectations and driving sustainability outcomes. These questions collectively aim to provide comprehensive insights into the adoption, effectiveness, and contextual factors of sustainable business models.

The objectives of the study are to assess the adoption of innovative business models, identify the drivers and barriers to their adoption, examine their impact on organizational performance, explore variations based on organizational characteristics, understand the role of stakeholder engagement, provide practical recommendations for businesses, and enhance strategic decision-making. These objectives collectively guide the research, fostering a comprehensive investigation into the intricate dynamics between innovative business models and sustainability, with the ultimate aim of contributing both theoretical knowledge and practical insights.

These research questions and hypotheses aim to guide the empirical analysis, allowing for a systematic investigation into the intricate dynamics between innovative business models and sustainability across a diverse array of organizational settings.

Data and Methodology

Research Model

In the study's conceptual framework depicted in Figure 13.1, the independent variable is IBM, encompassing diverse approaches that organizations employ to create, deliver, and capture value with a focus on innovation and adaptability. The dependent variable is sustainability outcomes (SO), which refers to the environmental, social, and economic impacts resulting from the integration of innovative business models. stakeholder engagement (SE), serving as the mediating variable, represents the extent to which organizations involve and consider the interests of various stakeholders in developing and executing innovative business models. organizational characteristics (OC), including size, industry type, and geographical location, function as the moderating variables. The study explores several relationships: a direct relationship (H1) where IBM positively impacts SO, a mediating relationship (H2) where SE mediates the relationship between IBM and SO, and moderating relationships (H3) where OC moderates the relationship between IBM and SO.

This framework integrates insights from multiple theories to provide a comprehensive

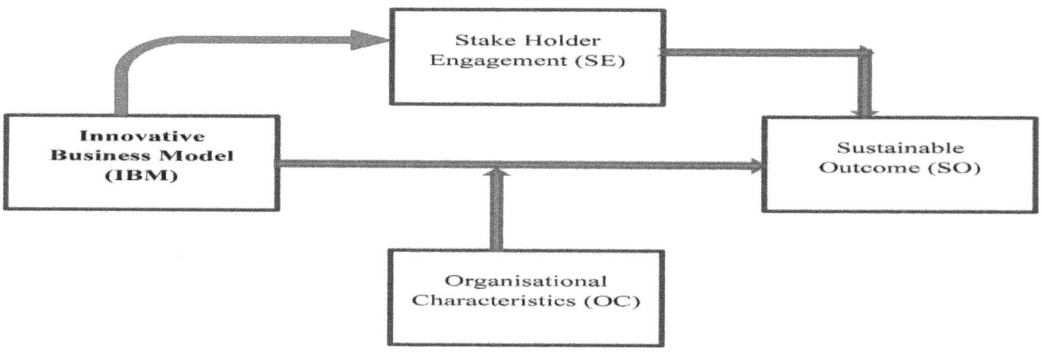

Figure 13.1 Research model conceptualized by the author
Source: Data analysis results

understanding of how IBM contributes to SO. Drawing from the RBV, it conceptualizes IBM as unique resources that enhance sustainability. Incorporating institutional theory, the framework highlights the role of SE in legitimizing and diffusing IBM within organizational practices. OC, serving as moderators, align with contingency theory, acknowledging that the IBM-SO relationship varies based on organizational context. Beyond its theoretical elegance, this framework offers practical applicability by illustrating how specific business models, through stakeholder engagement and organizational characteristics, contribute to sustainable outcomes. Grounded in established theories and tailored to the context of innovative business models and sustainability, it establishes a robust foundation for empirical analysis, facilitating the exploration of direct and contingent relationships to provide a multi-faceted understanding of achieving sustainability through innovative business practices.

This study employs a mixed-methods research design to explore the relationship between innovative business models and sustainability outcomes. It integrates quantitative and qualitative approaches to provide a comprehensive analysis. The study targets organizations across diverse industries in India that have adopted innovative business models with a sustainability focus, utilizing a stratified random sampling strategy based on industry type, organizational size, and geographical location. A minimum of 108 organizations, selected from the 'Fortune Global 500,' ensures statistical robustness. Quantitative data is collected through structured surveys and secondary financial data, while qualitative insights are gathered from in-depth interviews with key stakeholders. Data analysis involves descriptive and inferential statistics, such as regression and ANOVA, for quantitative data, and thematic analysis for qualitative data. Findings are triangulated to ensure a comprehensive understanding within the theoretical framework. Ethical considerations include obtaining institutional review board approval, informed consent, and maintaining confidentiality and anonymity throughout the research process

Findings and Discussions

From the (Table 13.1) Annexure-1, Data interpretation: mean scores for IBM1, IBM2, SE2, and OC1 are consistently high around 4.5, indicating a generally positive response or high ratings in these categories. Standard deviations and ranges are relatively small, suggesting low variability within each variable, with high precision in estimating mean values. Kurtosis values indicate a leptokurtic distribution for IBM1, IBM2, SE2, and OC1, while SE1 and SO have a mesokurtic distribution. Modes and medians are close to the mean for most variables, indicating a relatively symmetric distribution. Narrow confidence intervals suggest a high level of confidence in the accuracy of the mean estimates. Comparison across variables: IBM1, IBM2, SE2, and OC1 have identical means, standard errors, and ranges, indicating similarity in respondent opinions. SE1 and SO show uniqueness in skewness and kurtosis.

From the (Table 13.2), Correlation analysis reveals pronounced correlations between IBM1, IBM2, OC1, and OC2, with moderate relationships observed between SE1 and SE2. Perfect positive correlation between IBM1 and IBM2 suggests respondents who rated IBM1 high also tended to rate IBM2 high. Moderate positive correlation between IBM1 and OC2 indicates a relationship, while negative correlations between SE1 and OC2, and SE2 and SE1, suggest an inverse relationship. Limited linear relationships observed in most variables.

From the (Table 13.3), high correlation between independent variables and the dependent variable, with R Square value indicating the model explains 100% of the variance. ANOVA results suggest the

Table 13.2 Correlation analysis.

	IBM1	*IBM2*	*SE1*	*SE2*	*OC1*	*OC2*	*SO*
IBM1	1						
IBM2	-1	1					
SE1	-0.577	0.57735	1				
SE2	0	0	-0.57735	1			
OC1	0	0	0.57735	-1	1		
OC2	0.5774	-0.57735	-1	0.57735	-0.57735	1	
SO	0.5774	-0.57735	-0.33333	-0.57735	0.57735	0.333333	1

Source: Data analysis results

regression model is statistically significant, with all alternate hypotheses accepted, and precise estimation of coefficients indicated by extremely small standard errors.

The interpretation of the data reveals compelling insights into the relationships between variables. The high mean scores for IBM1, IBM2, SE2, and OC1 indicate strong favorability towards innovative business models, stakeholder engagement, and organizational characteristics, supported by low variability within each variable. Identical means, standard errors, and ranges across IBM1, IBM2, SE2, and OC1 underscore consistency in respondent opinions. Pronounced correlations between IBM1, IBM2, OC1, and OC2 suggest interdependencies in organizational practices, while moderate correlations imply potential relationships between innovative business models and certain organizational characteristics. The regression analysis indicates a robust relationship between the adoption of innovative business models, stakeholder engagement practices, organizational characteristics, and sustainability outcomes, further supported by ANOVA results. These findings underscore the significant roles played by stakeholder engagement and organizational characteristics in shaping sustainability initiatives, providing valuable insights for sustainable business practices.

The data interpretation and findings of this study align closely with the outlined objectives, offering a comprehensive understanding of the adoption and impact of innovative business models on sustainability outcomes. The strong inclination towards innovative business models among respondents underscores their adoption. Correlations between variables such as innovative business models, stakeholder engagement, and organizational characteristics help identify drivers and barriers to adoption. Regression analysis reveals the impact of these models on organizational performance, while comparisons across different variables highlight variations based on organizational characteristics. The significant role of stakeholder engagement is emphasized through its correlations with organizational characteristics, which indirectly suggests practical recommendations for businesses. These insights collectively contribute to enhancing strategic decision-making in sustainability initiatives. The study, however, is limited by its sample size and focus on Indian organizations, and its inability to capture longitudinal changes and causality. Nonetheless, the findings provide strategic insights for leveraging innovative business models for sustainable competitive advantage and contribute to sustainable development goals. Future research could explore the long-term effects

Table 13.3 Regression analysis.

Summary output	
Regression statistics	
Multiple R	1
R Square	1
Adjusted R Square	0.971154
Standard error	3.17E-16
Observations	108

ANOVA

	df	SS	MS	F	Significance F
Regression	6	5.0625	0.84375	1.68E+31	0
Residual	104	1.05E-29	1.01E-31		
Total	110	5.0625			

	Coefficients	Standard error	t Stat	P-value	Lower 95%	Upper 95%	Lower 95.0%	Upper 95.0%
Intercept	9	6.49E-16	1.39E+16	0	9	9	9	9
IBM1	0	0	65535	0	0	0	0	0
IBM2	-2.6E-16	8.63E-17	-2.97113	0	-4.3E-16	-8.5E-17	-4.3E-16	-8.5E-17
SE1	-0.5	1.22E-16	-4.1E+15	0	-0.5	-0.5	-0.5	-0.5
SE2	-0.5	8.63E-17	-5.8E+15	0	-0.5	-0.5	-0.5	-0.5
OC1	0	0	65535	0	0	0	0	0
OC2	0	0	65535	0	0	0	0	0

Source: Data analysis results

of these models on sustainability, their effectiveness across different industries and regions, and the motivations and challenges in adopting them for sustainability.

Conclusion

This study shows how important innovative business models are for achieving sustainability. It focuses on the roles of stakeholder engagement and organizational characteristics in making these models successful. The research found that companies using innovative business models perform better in terms of environmental, social, and economic outcomes. This conclusion is supported by strong data correlations and regression results. Although the study had some limitations, such as a small sample size and a focus on a specific geographic area, it still provides valuable insights for businesses looking to combine innovation with sustainability. The study aligns well with its objectives and offers practical advice for businesses navigating today's complex commercial environment. It shows how innovation and sustainability are connected, adding to the conversation about responsible business practices. A key takeaway is the importance of engaging with stakeholders and having supportive organizational characteristics to drive sustainability efforts. The research suggests that future studies should look at these dynamics in different contexts and over longer periods to build on these findings. Overall, the study gives businesses actionable insights and strategic tools to effectively adopt sustainable strategies. This helps in making informed decisions and planning strategically for a sustainable future. By understanding the link between innovation and sustainability, businesses can better foster innovation while also caring for the environment and society. This approach contributes to broader sustainable development goals and encourages more responsible business behavior. The study also highlights practical steps businesses can take to overcome challenges in adopting sustainable practices. These steps include improving stakeholder engagement and fostering an organizational culture that supports sustainability. By following these recommendations, businesses can achieve long-term success and make a positive impact on society. This research provides a comprehensive understanding of how innovative business models can drive sustainability. It equips businesses with the knowledge and tools needed to navigate the complexities of modern commerce, promoting a more sustainable and resilient future. The findings emphasize the importance of combining innovation with sustainability to achieve not only business success but also a positive societal impact.

References

Barney, J. (1991). Firm resources and sustained competitive advantage. *Journal of Management*, 17(1), 99–120.

Chen, Lin, and Yu Wang. 2023. "Sustainable Supply Chain Practices in the Fashion Industry." Journal of Fashion Marketing and Management 37 (2): 185–207.

Chesbrough, H. (2010). Open Innovation: The New Imperative for Creating and Profiting from Technology. Harvard Business Press.

Das, S. C., and Das, S. (2020). COVID-19 crisis: a study on the missing woman in benefit transfer through PMJDY. *Horizon*, 2, 53–58.

Das, S. L., Patnaik, B. C. M., Satpathy, I., and Das, S. C. (2022).A mathematical index model analysis of working capital management efficiency in the Indian private healthcare sector. *Journal of Pharmaceutical Negative Results* 13:1216–1221.

Das, S. L., Patnaik, B. C. M., Satpathy, I., and Das, S. C. (2022).Increase in the prevalence of mental health problems due to the COVID-19 pandemic and its impact on the work performance of private hospital supervisors in India. *Journal of Positive School Psychology* 6:6744 –6751.

Das, S. L., Patnaik, B. C. M., Satpathy, I., and Das, S. C. (2022). Working capital management and financial health of private hospitals of India after COVID-19 pandemic. *International Journal of Health Sciences* 6:1911–1922.

DiMaggio, P. J., and Powell, W. W. (1983). The iron cage revisited: institutional isomorphism and collective rationality in organizational fields. *American Sociological Review*, 48(2), 147–160.

Elkington, John. Cannibals with Forks: The Triple Bottom Line of 21st Century Business. Oxford: Capstone Publishing, 1997.

Garcia, Maria, Juan Perez, and Lucia Martinez. 2022. "Regenerative Agriculture as a Sustainable Business Model: Enhancing Soil Health and Biodiversity." Agricultural Systems 95(3): 321–334.

Geissdoerfer, M., Savaget, P., Bocken, N. M., and Hultink, E. J. (2017). The circular economy – a new sustainability paradigm? *Journal of Cleaner Production*, 143, 757–768.

Johnson, Mark, and Alice Lee. 2024. "Platform-Based Business Models in the Renewable Energy Sector." Energy Policy 78 (2): 233–246.

Porter, M. E., and Kramer, M. R. (2011). Creating shared value. *Harvard Business Review*, 89(1-2), 62–77.

Schiederig, T., Tietze, F., and Herstatt, C. (2012). Green innovation in technology and innovation management—an exploratory literature review. *R&D Management*, 42(2), 180–192.

Smith, John, Emily Davis, and Michael Brown. 2023. "Circular Economy Business Models: An Empirical Analysis." Journal of Sustainable Business 45 (1): 101–123.

Teece, D. J. (2010). Business models, business strategy and innovation. *Long Range Planning*, 43(2-3), 172–194.

Annexure-1

Table 13.1 Descriptive statistics.

IBM1		*IBM2*		*SE1*		*SE2*		*OC1*		*OC2*		*SO*	
Mean	4.5	Mean	4.5	Mean	4.25	Mean	4.5	Mean	4.5	Mean	4.75	Mean	4.625
Standard error	0.048337	Standard error	0.048337	Standard Error	0.041861	Standard error	0.048337	Standard error	0.048337	Standard error	0.041861	Standard error	0.02093
Median	4.5	Median	4.5	Median	4	Median	4.5	Median	4.5	Median	5	Median	4.5
Mode	4	Mode	5	Mode	4	Mode	4	Mode	5	Mode	5	Mode	4.5
Standard deviation	0.502331	Standard deviation	0.502331	Standard deviation	0.435031	Standard Deviation	0.502331	Standard Deviation	0.502331	Standard deviation	0.435031	Standard deviation	0.217516
Sample variance	0.252336	Sample variance	0.252336	Sample variance	0.189252	Sample variance	0.252336	Sample variance	0.252336	Sample variance	0.189252	Sample variance	0.047313
Kurtosis	-2.0381	Kurtosis	-2.0381	Kurtosis	-0.64091	Kurtosis	-2.0381	Kurtosis	-2.0381	Kurtosis	-0.64091	Kurtosis	-0.64091
Skewness	0	Skewness	0	Skewness	1.171028	Skewness	0	Skewness	0	Skewness	-1.17103	Skewness	1.171028
Range	1	Range	1	Range	1	Range	1	Range	1	Range	1	Range	0.5
Minimum	4	Minimum	4	Minimum	4	Minimum	4	Minimum	4	Minimum	4	Minimum	4.5
Maximum	5	Maximum	5	Maximum	5	Maximum	5	Maximum	5	Maximum	5	Maximum	5
Sum	486	Sum	486	Sum	459	Sum	486	Sum	486	Sum	513	Sum	499.5
Count	108	Count	108	Count	108	Count	108	Count	108	Count	108	Count	108
Confidence level (95.0%)	0.095822	Confidence level (95.0%)	0.095822	Confidence level (95.0%)	0.082984	Confidence level (95.0%)	0.095822	Confidence level (95.0%)	0.095822	Confidence level (95.0%)	0.082984	Confidence level (95.0%)	0.041492

Source: Data analysis results

14 Consumer perception toward eco-friendly products: a comparative study between rural and urban consumers in West Bengal

Sriparna Guha[a]

Narula Institute of Technology, Kolkata, West Bengal, India

Abstract

Now a days green goods have become popular among the consumers because these are more durable, recyclable, non-toxic, and made of materials that are either decomposable or recyclable. Consumers are becoming more concerned with the environment and looking for green products. Aim of this paper is to understand the perception of the respondents regarding the use of eco-friendly products and their benefits. In such conceptual background, the present empirical research tries to identify important factors that influence consumer purchase behavior for green products and green marketing, using factor analysis. The findings are also helpful for the marketers to put importance on the eco-friendly products and focuses on effective marketing strategy. Most of the FMCG producing companies are attracted to green marketing as they are interested in becoming successful in such diligent market. Green marketing has every potential to become a comprehensive pace-setter. In this regard, not only customer satisfaction, but also developing a better sense of ethical responsibility towards natural environment, is the need of the hour. In this study, 200 respondents were selected from three districts, namely - North 24 Parganas, Kolkata and Nadia, by using non-probability convenience sampling. Consumer's attitude towards environment, Product packaging, and product benefit emerged as three significant factors which have influence on consumer behavior for green purchase. Hence, the current study also possesses the importance of using eco-friendly products for fulfilling sustainable development.

Keywords: Consumer attitude, consumer perception, eco-friendly product, environmental product, green marketing

Introduction

The present society is seriously concerned about the environment as the degradation of the environment is a result of economic activity-led growth and development throughout the world. As a result, the intellectual world is more concerned about sustainable development and different goals to achieve sustainable development that started through the Kyoto Protocol. Hence, governments of both developed and developing nations are more stringent in taking serious initiatives to protect the environment.

The simple fact that needs to be remembered in the present scenario concerns the impact of production, distribution, and consumption of products on the natural environment, and many research efforts over the years to create what is known as "green" or environment-friendly products have already been a serious topic of discussion. The term green product is used to designate goods that are minimally harmful to the natural environment, which, talks about both the manufacturing process (resource use and pollutant emission) and the nature of such products (bio-degradable) along with their distribution at various levels.

[a]sriparna.guha@nit.ac.in

DOI: 10.1201/9781003564447-14

The term "green advertising" has its initiation from "green product". It is well known that the advertisement of any product is crucial in shaping the consumption pattern Khedkar (2015). Several researchers have estimated that appropriate advertising practices can guide habits and behavioral patterns of an individual's daily life in such a way that they tend to consume certain products. Hence, advertising efforts and policies engrossed in the creation of such shifts toward the purchase of green products are mentioned as green advertising. Polonsky and Everitt, during the late 1980s and early 1990s, coined the term green marketing.

As a policy of implementation of sustainable development, businesses have started to respond to environmental barriers through the usage of green marketing strategies Cherian and Jacob (2012). It was established that since the 1980s, ecological concerns like pollution, climate change, the greenhouse effect, and global warming, have a direct linkage to the production from industrial sectors. Presently, most of the companies and organizations have adopted and implemented green development ideas. Such policies are evident in the available sustainability reports. Thus, it is obvious that trades in most industries contemplate aspects of "environmental sustainability" during the process of designing goods and services. One important element in this regard is sustainable marketing.

Sustainable marketing comprises the development and preferment of goods and services that take care of customer needs through the maintenance of product quality, efficiency, price, and suitability without imposing any negative impact on the natural environment, society, or economy. It deploys high-end techniques that promote sustainable lifestyles to the ultimate consumers.

Technology has always played a pivotal role in creating value for the customer. Data being a significant factor in improving the marketing environment has always taken the help of technologies in producing data. The growth of business activities through customer satisfaction is essentially possible through the data generated in the process. The new generation application of social networks and mobile applications act as "social sensors" and capture people's emotions. The data generated on people's views and requirements are cost-effective and important for strategizing marketing activities. These, in turn, are needed for real-time analysis, and big data management in marketing. In this world of globalization, coupled with climate change, the purchasing behavior of individual consumers is required to impose environment-friendly purchasing behavior. The next section deals with the existing literature on different issues of green marketing.

Literature Review

In the words of the American Marketing Association, green retailing is the vending of merchandise that guarantees environmental-friendliness (i.e., these products are biodegradable) along with the exertions occupied by a company for production, packaging, branding, and modification of the finished product, for the deployment of technologies essential for ecological sustainability. Green merchandising is a barter procedure of fulfilling humanitarian desires and wishes in which all actions must be deliberate with minimalistic harm to the biosphere, yet with the receptivity to satisfy our needs (Polonsky, 1994). The overall view of green marketing implied the consideration and authorization for the adoption of environment-friendly strategies by consumers, taking into account the entire chain of events the firm enlists in, due to the very nature of natural issues (Hess and Timen, 2008). The growing concern of buyers for the survival of the biosphere, coupled with their protection, drives along the demand for green products, which motivates, and oversees a significant expansion in the ecological performances of many companies. Consumers are the most prominent factor that can take care of the procedures towards green marketing (Sachdev, 2011).

Consumers are not fully guilty of environmental protection, and may sometimes allow industries and governments to assume too much of their accountability (Maheshwari and Malhotra, 2011). Chen and Chai (2010) pointed out that customer-beneficial approaches regarding ecological protection did not speed up consumer attitudes toward green products. Ahmed and Rajendran (2014) have portrayed this in their study. They investigated that buying patterns could be altered with a certain degree of consumer awareness, but marketers and marketing should focus on educating their consumers on the increasing help that green products could provide.

A survey carried out by Chang and Fong (2010) on buyers experienced in purchasing green merchandise noted the identification of an affirmative association between the quality of the product and its subsequent effect on the satisfaction of the respective buyers. It was also explored by their study that a green corporate image can point toward the loyalty of green consumers which leads to consumer satisfaction. Hamid and Jameel (2019) in their study of Ayurveda in Kerala concluded that consumer's action has proportional impact on the natural environment. Another study based on the perception and the awareness among the consumer about green products has been found that the Indian youngsters were having demand for green products without any prior knowledge about it. In another study conducted by Gan et al. (2008) it has been seen that environmental consciousness has positive impact on the purchase decision of a consumer, however, brand consciousness has a negative influence on a consumer's purchasing decision of green products. The study also found that higher prices accompanied by brand unfamiliarity reduce the probability of purchasing green products. In search of the factors influencing green marketing in Bangladesh, Rayhan et al. (2014) shows that ethical marketing is the significantly important steps that can be taken to save nature. Yu et al. (2021), in their study on hotel industry, have concluded that the brand images and behaviours of the customers lead to eco-friendly hotel activities. The study shows that hotel's eco-friendly activities have a positive influence on hotel performance.

Another study concluded that though consumers are fond of purchasing green products, even at higher prices, seller should also put emphasis on consumer benefits by making green products economical compared to their non-green counterparts. Further, a study conducted by Nosratabadi et al. (2019) focuses on state-of-the-art sustainable business models which are essential for the development of future research.

From the above discussion, it has been found that green marketing is very significant from the environmental point of view and every company also emphasizes producing green products for the customers. The young generation can talk about green products but does not have the knowledge about green products and it is necessary to understand people's perception regarding green marketing and the differences in the perception between rural and urban populations Akter (2012).

From the above discussion it has been found that green marketing is significantly relevant from the environmental point of view and every company also emphasized on producing green products. The young generation can talk about green product but not have the knowledge about green product and it is necessary to understand the people's perception regarding green marketing and differences of the perception between rural and urban population on green marketing. Considering the variation in customers, in terms of area of residence (rural or urban), educational status, the following research questions are important.

1) Does environmental knowledge have any impact on purchasing of eco-friendly products?
2) Does residential location have impact on eco-friendly product purchase?
3) The hypotheses are:
   a) H_{01}: *There is no relationship between sources of information about green product by gender*
   b) H_{02}: *There is no significant relationship between opinion of the respondents toward eco-friendly product*
   c) H_{03}: *There is no relationship between perceptions of the eco-friendly product with income level.*

Methodology

The study has been conducted in three districts of southern West Bengal, namely North 24 Paraganas, Kolkata and Nadia. The respondents from Kolkata represent urban respondents, whereas the respondents from South 24 Paraganas and Nadia represented rural residents. Based on the requirement of the study, convenience sampling methods have been adopted for the identification of samples. In the present study, altogether, 200 samples were identified from both urban and rural areas of the above-mentioned three districts. For the purpose of data collection, a structured questionnaire was developed, with socio-economic components along with objective-oriented variables, as the research questions were pointing to the probable relationship between location and perception regarding green marketing, education and perception regarding green marketing and gender and perception regarding green marketing. The dependent variables have been considered as environmental knowledge. This environmental knowledge (EK) has been measured in terms of five point Likert Scale. The data has been collected using the above-mentioned questionnaire through personal interview. The data analysis and presentation has been structured using direst and cross tabulation. Besides, to identify and measure the variability, ANOVA and Chi-square tests were conducted to analyze data. The degree of association between area of residence and environmental awareness has been captured through correlation analysis.

Data Analysis and Interpretation

The present section concentrates on data analysis and interprets the results that are important to answer the research questions or validates the hypotheses mentioned above. The section starts with Demographic profile of the respondents.

Average age of the respondents is 36.58 years. It is to be noted that for the present study the target population was the younger section of the population

Table 14.1 depicts that majority of the respondents that is 30.5% belongs to the income group between Rs.25, 000 to Rs.50, 000. Besides, 27% belongs to less than Rs. 1 lakh income group. However, around 16%, 13% and 13.5% of the respondents belong to less than Rs. 10,000, Less than Rs. 25,000 and more than Rs. 100000 income group.

Table 14.1 Income distribution of the respondents.

Income	No. of respondents	Percentage
<10,000	32	16
<25,000	26	13
<50,000	61	30.5
< 1,00,000	54	27
>1,00,000	27	13.5
Total	200	100

Source: Primary data, 2023

From Table 14.2, it has been found that majority of the respondents that is 29.5% have achieved the education up to graduation, 28% have other educational qualifications, 24.5% have completed post-graduation. On the other hand, only 135 have completed secondary education and only 5% of the respondents have completed Higher secondary level of education. Hence, Table 14.2 shows that majority of the respondents have achieved higher education, which portray higher level of awareness toward green marketing.

Table 14.3 shows that majority of the male respondents have received the information from shopkeepers and newspapers and female respondents received information from friends and shopkeepers. The above table shows that both friends and shopkeepers are significant source of information regarding green product, compared to newspaper.

The correlation between income and EK1, it has been found it is (-0.0134)

The correlation between area of residents and EK1 is 0.06606.

The correlation between area of residents and EK2 is 0.242996

The correlation between area of residents and EK3 is (-0.31223)

The correlation between area of residents and EK4 is (-0.31181)

The correlation between area of residents and Education is 0.230411

From the above table H_{11} is accepted so, there is a significant difference between information sources about the green product and gender.

Table 14.4 shows that the majority of the respondents, 51% never purchase eco-friendly products.49% of the respondents said they have purchased eco-friendly products in the past. When they were asked about type of the product it was found that three types of products they had purchased, like plates, jute bags, and toys.

H_{22} is accepted. There is a significant relationship between influences of the green products and with area of residence.

The Calculated χ^2 =5.648 with 4 degrees of freedom.

The table value of χ^2 = 11.14 with 4 degrees of freedom at 0.025 level.

Table 14.2 Education distribution of the respondents.

Education	No. of respondents	Percentage
Upto secondary	26	13
Upto HS	10	5
Upto graduation	59	29.5
Postgraduate	49	24.5
Others	56	28
Total	200	100

Source: Primary data, 2023

Table 14.3 Relationship between sources of information about green product by gender.

Gender	Male	Female	Total
Sources of information about green product			
Friends	14	38	52
Shopkeepers	25	26	51
Newspaper	20	12	32
Others	19	17	36
Total	93	107	200

Source: Primary data, 2023

Table 14.4 ANOVA Table for variation between information sources on gender wise green product.

Summary	Count	Sum	Average	Variance		
Friends	3	104	34.66667	369.3333		
Shopkeepers	3	102	34	217		
Newspaper	3	64	21.33333	101.3333		
Others	3	72	24	109		
15	4	78	19.5	20.33333		
14	4	93	23.25	130.25		
29	4	171	42.75	104.9167		
ANOVA						
Source of variation	SS	df	MS	F	P-value	F crit
Rows	419.6667	3	139.8889	2.41999	0.164376	4.757063
Columns	1246.5	2	623.25	10.78184	0.010314	5.143253
Error	346.8333	6	57.80556			
Total	2013	11				

Source: Primary data, 2023

Table 14.5 Opinion of the respondents toward eco-friendly product purchase.

Opinion	Respondents	Percentage
Yes	97	49
No	103	51
Total	200	100

Source: Primary data, 2023

Table 14.6 Factors influenced the decision to buy eco-friendly products by area of residence.

Factors influenced the decision to buy an eco-friendly product	Urban	Rural	Total
Self	16	22	38
Peers	18	16	34
Packaging	56	46	102
Hygiene	4	4	8
Place	3	7	10
Promotion	3	5	8
Total	100	100	200

Source: Primary data, 2023

Therefore, H_{03} is accepted. so, there is no relationship between the opinions of the respondents regarding awareness of eco-friendly products with income.

Table 14.7 ANOVA Table indicating variation between area of residence wise decision of buying green products.

ANOVA: Two-factor without replication

Summary	**Count**	**Sum**	**Average**	**Variance**
Self	3	76	25.33333	129.3333
Peers	3	68	22.66667	97.33333
Packaging	3	204	68	892
Hygiene	3	16	5.333333	5.333333
Place	3	20	6.666667	12.33333
Promotion	3	16	5.333333	6.333333
Total	3	400	133.3333	3333.333
Urban	7	200	28.57143	1339.286
Rural	7	200	28.57143	1205.286
Total	7	400	57.14286	5062.476

ANOVA

Source of variation	SS	df	MS	F	P-value	F crit
Rows	40499.81	6	6749.968	15.75109	4.68E-05	2.99612
Columns	3809.524	2	1904.762	4.444774	0.035935	3.885294
Error	5142.476	12	428.5397			
Total	49451.81	20				

Source: Primary data, 2023

Table 14.8 Relationship between respondent opinions on eco-friendly product awareness to income.

Income	**Awareness of eco-friendly product**		**Total**
<Rs.10,000	12	20	32
<Rs.25,000	06	20	26
<Rs.50,000	20	41	61
<Rs.1,00.000	24	30	54
>Rs.1,00,000	7	20	27
Total	69	131	200

Source: Primary data, 2023

Conclusion

India, like many other responsible countries, faces rapidly incrementing environmental obstacles – people from all walks of life are indirectly or directly participating in eco-consciousness, a modern symbol of the successes of day-to-day transactions. There are many factors that influence the behavioral constructions of the eco-friendly products between consumers residing in rural and urban areas of a smart city such as Kolkata, all of which have been duly identified, analyzed and listed in this study. This paper contributes empirical research that uses factor-wise analysis of collected data from selected samples to identify important factors that influence consumer purchase behavior for green products and green marketing. Newspapers, magazines, and other forms of media are increasingly focusing on a product's potential environmental repercussions. This study depicts the actuality of a substantial contrast among male and female respondents on the basis of the persuasive factors noted by consumers when purchasing a green product Gan et al. (2014). The extant of a considerable dissimilarity between urban and rural consumers as to the green product marketing behavior. After a thorough analysis of the collected data, it has been found that the most important factor is the packaging of the products. Hence, the firms which are producing eco-friendly products should focus on the packaging of such products Fitzpatrick et al. (2012). Wadding is viewed upon as the fifth component alternatively the fifth 'P' of the marketing colloid, due to its importance in the initialization of the firm's communication with its buyers, in building a structured, constructive, productive relationship with them, thus influencing their purchasing decisions, promoting their creations and their distinctiveness from competitors.

References

Ahmed, R., and Rajendran, R. (2014). A study on consumer awareness of green products and its impact on green buying behaviour. *International Journal of Research*, 1(8), 1483–1493.

Akter, J. (2012). Consumer attitude towards green marketing in Bangladesh. *SA University Review*, 6(1), 157–166.

Chang, N., and Fong, C. (2010). Green product quality, green corporate image, green customer satisfaction, and green customer loyalty. *African Journal of Business Management*, 4(13), 2836–2844.

Chen, T. B., and Chai, L. T. (2010). Attitude towards the environment and green products: consumers' perspective. *Management Science and Engineering*, 4(2), 27–39.

Cherian, J., and Jacob, J. (2012). Green marketing: a study of consumers' attitude towards environment friendly products. *Asian Social Science*, 8(12), 117–126.

Fitzpatrick, L., Verghese, K., and Lewis, H. (2012). Developing the strategy. In Verghese, K., Lewis, H., and Fitzpatrick, L. (Eds.), Packaging for Sustainability. London: Springer London, pp. 1–40.

Gan, C., Wee, H. Y., Ozanne, L., and Kao, T. H. (2008). Consumers' purchasing behaviour towards green products in New Zealand. *Innovative Marketing*, 4(1), 93–102.

Gan, C., Zhiyou, C., Tran, M. C., Cohen, D. A., and Xiangxiang, W. (2014). Consumer attitudes toward the purchase of organic products in China. *Faculty of Agribusiness and Commerce,* Lincoln. University: Paper no. 15.

Hamid, S., and Jameel, S. T. (2019). A study of green marketing practices in the selected ayurvedic resorts of Kerala. In Global Developments in Healthcare and Medical Tourism. Hershey, PA, USA: IGI Global.

Hess, E. and Timen, P. (2008). Environmental Friendliness - A Marketing Strategy, Master Thesis within Business Administration, *Internationella Handelshogskolan Hogskolan Jonkoping*, 22(5), 385–398.

Khedkar, E. B. (2015). Green marketing- concept, awareness and linkage with consumer purchase decision. *International Journal of Management Sciences and Business Research*, 4(1), 14–18.

Kumar, S, Garg. R and Makkar A. (2012). Consumer Awareness and Perception Towards Green Products: A Study of Youngsters in India, International Journal of Marketing & Business Communication, 1(4), 35–43.

Maheshwari, A., and Malhotra, G. (2011). Green marketing: a study on Indian youth. *International Journal of Management and Strategy*, 2(3), 1–15.

Nosratabadi, S., Mosavi, A., Shamshirband, S., Zavadskas, E. K., Rakotonirainy, A., and Chau, K. W. (2019). Sustainable business models: a review. *Sustainability*, 11, 1663.

Polonsky, M.J. (1994) An Introduction to Green Marketing. Electronic Green Journal, 1, 2–3.

Rayhan, A. I., Khan, A. I., and Islam, M. T. (2014). Factors affecting green marketing in Bangladesh. *Global Journal of Management and Business Research: E Marketing*, 14(1), 23–40.

Saschdev, S. (2011). Eco-friendly products and consumer perception. *International Jour-*

nal of Multidisciplinary Research, 1(5), 1–12.

Yu, J., Park, J., Lee, K., and Han, H. (2021). Can environmentally sustainable development and green innovation of hotels trigger the formation of a positive brand and price premium? *International Journal of Environment Research Public Health*, 18(6), 3275.

15 Corporate performance and sustainability: the role of audit committees, board composition, and CEO duality

Saroj Kumar Routray[1,a] *and Sabat Kumar Digal*[2,b]
[1]School of Management, KIIT Deemed to be University, Bhubaneswar, India
[2]P.G. Department of Commerce, Ramadevi Women's University, Bhubaneswar, India

Abstract

This research aims to explore how certain governance factors of board within companies influence their performance in the Indian capital market. Precisely, it studies the impact of board composition, audit committee independence, and CEO duality, on accounting and stock market performance metrics. Using panel data regression analysis, the study analyses 123 enterprises of the BSE 200 index for a thirteen-year period from 2007-2019. The sample excludes banking and financial service companies regulated by the Reserve Bank of India (RBI). The analysis reveals significant relationships between company's performance and its governance factors. Specifically, certain governance practices, such as audit committee independence and board composition are found to positively influence both accounting and stock market performance metrics. The outcomes of the current study have both policy and managerial implications for policy maker's regulators and corporate leaders seeking to enhance corporate governance standards in India. Implementing these governance practices may lead to improved investor confidence, enhanced shareholders value, and ultimately, sustainable long- term performance. The current study offers a novel contribution to the prevailing literature by providing an empirical understanding of the association between businesses' performance and the corporate governance factors in the Indian context. This research adds to the knowledge on how governance practices impact a firm's financial performance thereby offering insights into potential avenues for further research and policy development.

Keywords: Audit committee, board composition, CEO duality, corporate governance, firm performance

Introduction

The corporate governance reform began in 1991 with the formation of the Cadbury Committee, led by Sir Adrian Cadbury. Initiated by the financial reporting council (FRC), London, the committee aimed to improve governance quality in companies. It recommended a best practice code for the corporate governance (CG) becoming a keystone for good governance in the UK and beyond. Subsequently, the Greenbury and Hampel committees were established with five years of the Cadbury Committees formation. These committees emphasized on enhancing transparency and reliability in accounting disclosures and increasing management's responsibility towards shareholders. A major focus was board composition, with suggestions for an increased ratio of independent directors and establishment of board sub-committees to address important corporate issues. The Cadbury Committee report was a milestone corporate governance reform history. Key recommendations included greater independence

[a]routray.finance@gmail.com, [b]sabatdigal@gmail.com

DOI: 10.1201/9781003564447-15

in the board of directors, separations of board chairmanship, and CEO positions. It suggested for having more independent directors on the board to influence decision-making and better manage agency issues. The committee also proposed to form an audit committee with at least three independent directors and suggested to have all independent directors if the committee is only consisting of three members, The audit committees primary responsibility is to supervise the firm's finance reporting and internal audit processes, ensuring the statutory auditor's independence. The overall objective was to enhance accounting disclosure reliability and transparency, reducing the scope for deliberate accounting misstatements.

The committee recommended a clear division of responsibility between the CEO and the chairman of the board to ensure an equitable distribution of authority and responsibility between these two top positions. While favouring separate individuals for these roles, the committee also suggested that if one person holds both positions, the board should have higher independence to maintain a balanced power dynamic. More independent directors on the board could enhance accountability in management, even if the CEO and chair are the same person. The Cadbury Committee offered numerous governance suggestions, but our brief discussion focuses on these three aspects relevant to our research.

Two decades from year 1990 to 2010 saw many corporate frauds across the countries. Many of them are the accounting manipulations to tamper and reshape the financial statements. Misleading the shareholders and investors was primarily the objective of all these intentional and deliberate attempts on the part of the management. Highly reputed companies like Enron, WorldCom, Global Crossing, Adelphia, Arthur Andersen, Tyco, Satyam, were all involved in the misrepresentation of accounting data either using the weakness of GAAP or using some unlawful practices. This had completely shaken the trust and confidence of investors. Many countries wanted to control (the excessive freedom given to the management and make them more responsible and accountable to their investors, so they brought many regulatory changes. This process started after nineties and expedited in the new century. Another significant milestone in corporate governance reform was the enactment of the Sarbanes-Oxley Act in the US in year 2002. The primary aim of the Act was to enhance the reliability of audit processes and accounting disclosures. Under the Act, the CEO of the company and also the CFO, both are required to certify the company's financial reports, making them accountable for the accuracy of these reports and answerable to shareholders for any misleading accounting practices by the management.

India was also watching the progress happening worldwide. It went with many such reforms keeping the corporate environment of the country in mind. The Ministry of Company Affairs and SEBI (Security Exchange Board of India) formed committees under the chairmanship of some of the prominent business leaders of the country with the objective of regulatory improvisation. Kumar Managalam Birla Committee, Narayan Murthy Committee, and Naresh Chandra Committee, are few of the committees to name, are formed. In 2005, based on the report of the Kumar Managalam Birla Committee, SEBI added Clause 49 of the listing agreement. This clause asked to give various mandatory and voluntary disclosures in the quarterly compliance reports, which are to be filed with the stock exchange in a prescribed format. These reports were to be signed by the CEO and CFO of the companies, making them responsible for what was reported. The companies were now required to have an additional section as a corporate governance section in the financial report of the company.

The committee recommended for adding a clause to strengthen the effect of the independent directors on the decision-making of the board. The clause requires not less than fifty percent independent directors on the board if it is led by the company's CEO, with a concession to have fewer independent directors if the CEO does not serve as the board chair. The independent directors, in that case, would be at least 1/3rd of the overall board members. Board composition is considered as an instrument to bring more reliability to the process of the management function and decision making which ultimately reduces the scope of promoting the interest of promoters and majority shareholders and protects the interest of minority shareholders.

The committees also recommended forming sub-committees comprising of some board members to reduce the autocratic decisions of any one person related to appointment, audit, and remuneration of directors and auditors, etc. One of the recommendations mandates the companies to arrange an audit committee with a higher ratio of autonomous directors. The intention was to enhance the confidence of stakeholders in the integrity of the process of accounting disclosure and improve the internal control of corporate reporting.

Realizing the potential damage that could be done to the basic format of the corporate organization by not safeguarding the interest of the shareholders, almost all countries worldwide have started doing big reforms in corporate governance and forcing the companies to adhere to the rules made by them.

In this context, it becomes important to analyse the influence of corporate governance reforms on enterprise performance. In this paper, we have considered three important governance factors, board structure, CEO duality, and independence of audit committee and tried to study their impact on the financial performance of the company.

Literature Review and Hypothesis Formation

Over the last two decades, numerous studies have examined the impact of governance factors and practices on firm performance. Goel et al. (2015) investigated corporate governance practices and their effect on working capital efficiency, observing a significant positive impact leading to better firm performance. Hermalin and Weisbach (1991), Daily and Dalton (1994), and Beasley (1996) studied the influence of board structure on shareholder wealth and company performance. Other studies explored the association between performance and CEO duality, where the chairperson and CEO roles are held by one or separate individuals, and its effect on firm performance (Molz, 1988; Daily and Dalton, 1992). In Indian banks, Kumar and Sudesh(2019) found a statistically significant positive correlation between earnings per share (EPS), an indicator of firm performance, and corporate governance. Reddy and Akula (2011) found similar results, suggesting effective corporate governance reduces principal-agent problems, increases accountability, and enhances market performance. Some studies examined the impact of independent directors on various committees like audit, compensation, and nominating on firm performance (Boyd 1995; Dechow et al., 1996; Beasley, 1996; Shivdasani and Yermack, 1999). Venkatraman and Selvam (2014) studied the impact of ownership structure and governance mechanisms on business performance. Furthermore, the varied outcomes across studies underscore the need for further research, particularly considering evolving contexts over time and across different countries. These inconsistencies highlight the necessity for additional studies to comprehensively understand the underlying dynamics.

Board Composition

The mandatory composition of boards in many countries is based on the assumption that independent board members are truly independent, free from managerial influence, and better equipped to guide and control management. Independent directors are thought to monitor management more effectively due to their detachment from company management (Fama and Jensen, 1983a, b) and their extensive experience and proficiency in governance. Their ability to supervise and control management is also motivated by the desire to uphold their goodwill in the external resource market.

Empirical studies indicate that independent board members are more engaged in managerial oversight, thereby protecting shareholder interests (Baysinger and Butler, 1985; Fama and Jensen, 1983a, b). Nicholson and Kiel (2004) studied 348 listed Australian companies and discovered a positive correlation between board independence and company performance. Abidin et al. (2009) observed a similar association in their study on Malaysian firms. Daily and Dalton (1994) observed that a higher percentage of independent board members positively impacts firm performance, a finding echoed by Bijalwan and Madan (2013).

However, some studies suggest contrary findings regarding the association between board composition and performance. Bhagat and Black (2001) did not identify any association between the presence of independent directors and improved company performance, a result supported by Kota and Tomar (2010) and Routray and Bal (2016). Fariha et al. (2022) found an adverse inverse relationship between independent directors and company performance in Bangladeshi companies. Saat et al. (2011) studied 221 Malaysian listed companies also observed a negative link between board independence and performance. This raises the question of whether simply having a higher proportion of independent directors is sufficient for effective management monitoring, or if active participation in board activities is also required.

Klein et al. (2004, 2005) identified an adverse relationship between these variables in highly concentrated Canadian firms. Similarly, Fosberg (1989) observed an insignificant relation between the proportion of independent directors and the firm's accounting performance. Hermalin and Weisbach (1991) did not find any statistically meaningful relation between the composition of board and market-based performance, such as Tobin's Q. Existing literature does not support a clear association between board composition and performance, and there is no consensus among scholars on this relationship. As a result, the current study introduces the following hypothesis.

H1: There exists no association between composition of the board and performance of the firm.

CEO Duality

The agency theory advocates for splitting the positions of CEO and Chairperson of Board between two different individuals ensures the balance of power between the two positions and no single person has excessive power of decision making. On one side, the plan of action is initiated and implemented by the CEO of the company, and in the other side, the board chairperson sees that the board of directors functions properly in observing and supervising the CEO of the company. The agency theorists say that if these two roles are combined and headed by a single person (CEO duality), then the person becomes very powerful and the position of the board becomes weak and it fails to monitor the opportunism of managers, leading to earnings management, exploitation of minority shareholders' interest, corporate frauds and scandals, etc. A study carried out by Berg and Smith (1978) on Fortune 200 companies, reports a negative association of CEO duality with Return on Income (ROI) and finds no significant relationship with the ROE and stock price change. Rechner and Dalton (1991), taking a subset of Fortune five hundred companies found a negative association of CEO duality with a company's ROE, ROI, and profit margin. In their study of 212 Sri Lankan companies, Wijethilake and Ekanayake (2020) found consistent results. Their findings indicate that when the CEO holds additional informal power, CEO duality is linked to an adverse impact on firm performance.

Contradict to the agency theorists' view, stewardship theorists suggest that CEO duality stretches unity of direction and evades the role indistinctness between two positions (Anderson and Anthony, 1986; Donaldson and Davis, 1991) and that expedites the process of decision-making in a potential dynamic business environment leading to improved performance of the firm. The research suggests that in certain circumstances, duality improves the firm performance and in some other cases it is inversely related to firm performance (Boyd, 1995; Lam and Lee, 2008). Due to the lack of consistent findings among researchers, the current study proposed the null hypothesis as follows:

H2: No association between CEO duality and the company performance.

Audit Committee Composition

The prominence of audit committee in the process of corporate governance has gone up over a period of time. The reliability of financial statements rests very much on the integrity and impartiality of audit committee members as they play a crucial role in the process of internal control or audit of a firm. Proper internal control leads to transparency in reporting and results in the reduction of fraud. Two things are very important in this regard i.e., accounting and financial understanding of audit committee members and their independence from external or internal influence. The audit committee's independence had been emphasised by Cadbury Committee and many other corporate governance committees across the world. Most of the governance committees emphasised the committee with the majority of independent directors.

Research findings suggest that larger audit committees enhance the capability to monitor the accounting process (Mohd Saleh et al., 2007), and audit committee independence reduces instances of corporate fraud (Persons, 2005). Bansal and Sharma (2016) noted a positive link between audit committee independence and both accounting performance and Tobin's Q. Wild (1994) observed that companies with audit committees tend to experience a positive market reaction following their earnings reports and exhibit superior market performance. Alqatamin (2018) in their study observed similar association between both the variables.

However, there were also findings of studies suggesting no conclusive link between audit committees and firm performance (Vafeas and Theodorou, 1998; Klein, 1998; Weir and Laing, 2001). The impact of audit committees on the reliability of financial statements also appears mixed. Some research indicates that having an independent audit committee enhances the reliability of financial information (Carcello and Neal, 2003; Dechow et al., 1996; McMullen and Raghunandan, 1996). Conversely, Beasley (1996) found no improvement in the reliability of financial statements due to the presence of an independent audit committee. Furthermore, Carter et al. (2007) observed that a larger ratio of independent directors in the committee is negatively associated with firm performance.

Overall, while some research underscores the benefits of larger and independent audit committees in enhancing firm performance and financial reliability, other studies indicate no significant impact or even negative associations. These mixed results suggest for further research in the topic and go deeper into understanding the reasons behind such variation in findings. In the context of India, the present study tried to observe the chemistry between independence of the committee and the performance of firm and consequently, proposes the following hypothesis.

H3: There is no association between audit committee independence and performance of companies.

Data and Research Methodology

The BSE 200 companies of the Bombay Stock Exchange (BSE) have been taken for the study. The market capitalisation percentage of BSE 200 companies represents more than 80% of that of BSE-listed companies. The data have been collected for all 200 companies but for the final study, only 123 companies are considered. Banking companies are not considered because of their special regulatory requirement by RBI and a few more companies are not taken because of data unavailability. Hence the final sample size remained at 123 companies. The data are taken for a period of thirteen years from 2007 to 2019 and the total number of firm-year observations was 1599. The firm's accounting and market performance numbers were gathered from the CMIE database and the corporate governance-related statistics were collected from the company's Annual Reports.

The research investigated the dataset encompassing 123 firm observations over 13 years, employing panel data regression analysis. The panel unit root was examined, stationarity of data was established and there were no unit root issues.

The study observed the impact of board features like board composition, CEO duality, and composition of audit committee on company performance using Least square (LS) fixed-effects and random-effects models. Variations within cross-sectional units and individual units over a period of 13 years are examined through the fFixed-effect and random-effects tests (Gaur and Gaur, 2006). The hypothesis here is that the regression parameters remain the same over time and also do not change among different cross-sectional elements, and that increases the consistency of coefficient projections.

In our study, we have considered three control variables under the assumption that the age of companies, leverage (debt/equity), and asset size of the company also influence the performance of the firm. Therefore, the impact of predictor variables is studied while keeping them as control variables. Due to the significant variation in asset size among different firms, we performed a log transformation on asset size.

Firm Performance

Tobin's Q is considered by Kaldor (1966) as a metric for assessing capital market performance, while ROE and ROA are commonly employed to gauge accounting performance i.e. return on equity and return on asset. In the current study, the capital market performance of the firm is assessed taking the market-to-book value ratio, an alternative to Tobin's Q, further referred in the paper as MVBV. The total share price of outstanding equity shares at the end of the accounting year is taken as the market value and the book value of shares is shareholders' equity at the end of the accounting year. The base model is formed in the following manner:

Firm Performance = f (board composition, CEO duality, composition of audit committee, leverage, company age, company asset size)

The model with capital market performance (MVBV) as the dependent variable (Model-1),

$$MVBV = \alpha + \beta_1 \text{ Board\_Comp} + \beta_2 \text{CEO\_duality} + \beta_3 \text{ Comp\_Audit} + \beta_4 \text{ Lev} + \beta_5 \text{ LnAge} + \beta_6 \text{LnSize} + \varepsilon$$

Table 15.1 Operational definitions of variables.

Variables	Indicators	Operational definition
Predicted variable:		
Performance of firm	MVBV/ROA/ ROE	Market value to book value ratio/accounting profit
Explanatory variable:	Board_Comp	Independent directors' in the board to the total members
	CEO_duality	A dummy variable: 0, if one individual occupies both positions of chairman and CEO of the board, 1 otherwise.
	Comp_Audit	Independent directors to all directors in the Audit Committee
Control Variable:	Lev	Total borrowings/total assets
	LnAge	Natural logarithm of firm age
	LnSize	Natural logarithms of the assets value of firm

Source: Authors' own description

Table 15.2 Summary statistics.

	Number	Min.	Max.	Mean	Std. DV.
MVBV	1599	0.2	39.85	4.994	4.949428
ROE	1599	-283.49	791.74	24.84	33.87
ROA	1599	-41.85	127.77	11.94	9.83
Board_Comp	1599	0	0.857143	0.511	0.115249
CEO_duality	1599	0	1	0.307	0.461871
Comp_ Audit	1599	0	1	0.865	0.150545
Lev	1599	0	0.613546	0.173	0.159305
Firm age (years)	1599	0	117	44.150	24.28715
Firm size (Rs. Million)	1599	1210.4	3677440	193310.9	386070.3

Source: Authors' own description

The model with accounting performance (ROE) as the dependent variable (Model-2),

$ROE = \alpha + \beta_1 \text{ Board\_Comp} + \beta_2 \text{ CEO\_duality} + \beta_3 \text{ Comp\_Audit} + \beta_4 \text{ Lev} + \beta_5 \text{ LnAge} + \beta_6 \text{ LnSize} + \varepsilon$

The model with accounting performance (ROA) as the dependent variable (Model-3),

$ROA = \alpha + \beta_1 \text{ Board\_Comp} + \beta_2 \text{ CEO\_duality} + \beta_3 \text{ Comp\_Audit} + \beta_4 \text{ Lev} + \beta_5 \text{ LnAge} + \beta_6 \text{ LnSize} + \varepsilon$. Table 15.1.

Analysis and Findings

Descriptive Statistics

Table 15.2 reveals that the average market value to book value (MVBV) ratio is approximately 5, though it varies widely from 0.2 to about 40. The board composition ratio has a mean of 0.511, indicating that over 50% of board of directors are non-executive/independent. The ratio of independent members in the audit committee is notably high at 0.865, indicating that many companies have audit committees composed entirely of independent members. The average debt-to-asset ratio is below 20%, indicating that Indian companies generally have low debt levels in their capital structure. Additionally, the average asset size of the companies is around INR 19,331 crores, reflecting that the sample primarily includes large and medium-sized companies.

Table 15.3 Correlation matrix of variables.

	MVBV	ROE	ROA	Board_ Comp	Comp_ Audit	Leverage	Firm Age	Firm Size
MVBV	1							
ROE	0.442***	1						
ROA	0.440***	0.536***	1					
Board_Comp	-0.053*	-0.032	-0.044	1				
Comp_Audit	0.052*	0.030	0.000	0.314	1			
Lev	-0.30***	-0.20***	-0.49***	0.07**	0.092***	1		
Age	0.056*	0.045	-0.021	0.012	0.002	-0.022	1	
Firm-Size	-0.21***	-0.11***	-0.17***	-0.042	0.106***	0.144***	0.035	1

Source: Authors' own estimation

Correlation Analysis

Table 15.3 presents the key relationships among the variables MVBV shows a positive correlation with accounting return of the firm. Composition of board exhibits a negative correlation with MVBV at a higher significance level (10%). Audit team composition and MVBV shows a positive relationship. Board independence and composition of is having a positive correlation. Leverage (debt ratio) is not consistently related to all measures of company performance but shows a positive correlation with board composition and the audit committee independence. This suggests that higher leverage prompts firms to include more independent directors on both the board and audit committee, aiming to enhance organizational reliability and trustworthiness among the lenders. However, the negative relationship with company performance indicates that companies experiencing lower profitability or in their early stages of establishment tend to have higher debt levels. Firm age demonstrates a positive correlation with market performance, implying that older firms typically exhibit higher MVBV ratios, possibly due to the lower valuation of depreciated assets and properties. Firm size is negatively correlated with the MVBV ratio and board composition, while it shows a positive correlation with both audit committee composition and leverage.

Regression Analysis

The regression analysis is carried out to see whether the model fits the data and the independent variable is having any impact on the dependent variable. For panel data, we have carried out the regression with both fixed and random effect to see which model fits more. Referring the Table 15.4 and Table 15.5, it is observed that the results of fixed and random effects are significant with $p < 0.05$, hence Hausman test has been done to see which model is appropriate to predict the relationship. The Hausman test result shows 'p' value less than 0.05 (Table 15.6), hence the null hypothesis is not accepted. Therefore fixed effect is appropriate for the model.

The Fixed Effect model demonstrates the best fit concerning the dependent variables MVBV and ROA. Table 15.4 reveals that both board composition and CEO duality negatively influence the market value of shares. The significance levels for board composition and CEO duality are noted at $p < 0.01$ and $p < 0.5$, respectively, indicating that the market views higher independence in the board unfavourably. Similarly, the market prefers the consolidation of board

Table 15.4 OLS estimates of relationship (fixed-effect regression result).

	Model-1: MVBV	Model-2: ROE	Model-2: ROA						
	Coefficient	t-value	Significance (p)	Coefficient	t-value	Significance (p)	Coefficient	t-value	Significance (p)
Constant	11.980	7.973	0.000	40.512	3.690	0.000	24.111	8.521	0.000
Board_Comp	-2.500	-1.875	0.061	-9.502	-0.975	0.330	-3.173	-1.263	0.207
Comp_Audit	4.338	4.245	0.000	16.737	2.241	0.025	5.082	2.640	0.008
CEO_duality	-0.769	-2.372	0.018	-1.677	-0.708	0.479	0.625	1.024	0.306
Lev	-0.887	-8.374	0.000	-3.016	-3.898	0.000	-0.992	-4.974	0.000
LnAge	0.301	1.459	0.145	2.674	1.772	0.077	0.294	0.755	0.450
LnSize	-7.866	-8.464	0.000	-37.468	-5.518	0.000	-29.144	-16.651	0.000
Adj. R^2	0.177			0.062				0.265	
"F" value	31.23			10.309				50.457	
Significance	0.000			0.000				0	

Source: Authors' own estimation

Table 15.5 OLS estimates of relationship (random-effect).

Model	Model-I: MVBV	Model-II: ROE	Model-III: ROA						
	Coefficient	t-value	Significance (p)	Coefficient	t-value	Significance (p)	Coefficient	t-value	Significance (p)
Constant	13.612	6.441	0.000	49.203	3.617	0.000	35.82	8.609	0.000
Board_Comp	-0.190	-0.137	0.891	-13.237	-1.151	0.250	-5.216	-1.677	0.094
Comp_Audit	1.247	1.229	0.219	14.991	1.713	0.087	1.133	0.488	0.626
CEO_duality	-0.480	-0.999	0.318	-1.008	-0.335	0.738	1.932	2.028	0.043
Lev	-0.936	-5.633	0.000	-3.463	-3.533	0.000	-1.710	-5.428	0.000
LnAge	0.159	0.380	0.704	2.564	1.291	0.197	0.433	0.612	0.541
LnSize	-2.896	-2.595	0.010	-37.862	-4.548	0.000	-29.265	-12.14	0.000
Adj. R^2	0.039			0.040				0.162	
"F" value	6.699			6.910				28.044	
Significance	0.000			0.000				0.000	

Source: Authors' own estimation

Table 15.6 Hausman test result (for identifying best model fit).

Model	Independent variable	Chi-Sq. statistic	Probability (p-value)	Suitable model
1	MVBV	13.210	0.040	Fixed
2	ROE	8.678	0.193	Random
3	ROA	27.513	0.0001	Fixed

Source: Authors' own estimation

chair and CEO roles, rather than their division between two individuals. These findings suggest that such scenarios result in management decision delays, causing firms to miss opportunities and ultimately harming investor interests.

Regarding the composition of the audit committee, a statistically significant positive association is observed with both MVBV and ROA. It can be said that more independent members in audit committee reduces the likelihood of accounting manipulation and safeguards minority shareholder interests in the firm. Therefore the accounting performance of firm improves and positively boosts the confidence of the investors.

In terms of accounting performance, ROE, the random effect model fits best. Board composition and CEO duality show no statistically significant relationship with ROE. However, there exists a positive association of audit committee independence with firm performance at the $p < 0.1$ level.

Control variables, such as asset size and leverage, exhibit an adverse association with performance, both accounting and stock market return. This suggests that high leverage may stem from the firm's inability to generate sufficient cash flow from operations, leading to increased debt in the capital structure. Elevated debt levels amplify financial risk and raise capital costs, negatively impacting firm performance. Additionally, larger firms demonstrate a statistically significant negative relationship with firm performance, indicating reduced operational efficiency and flexibility in capitalizing on immediate opportunities. The age of the firm shows a positive relationship with firm performance, although this relationship is not statistically significant.

Conclusion

Over the past three decades, corporate governance related research has garnered significant attention, particularly in examining its variables' correlation with company performance. However, this research has been marked by inconclusive and inconsistent findings, exacerbated by differing regulatory frameworks and enforcement levels between developed and developing/emerging economies. Consequently, there's a critical need for further research in these economies to identify weaknesses and propose policy-level interventions. This study aims to evaluate the impact of key recommendations of various governance committees regarding board composition, separation of board chairmanship and CEO positions, and the increase in independent directors on audit committees, and their effects on company performance.

The analysis revealed two distinct company groups: those adhering solely to mandatory recommendations and those exceeding minimum requirements by appointing extra independent directors to both the board and audit committees. Findings regarding composition of board and CEO duality displayed inconsistency in their association with market and accounting performance. While both factors exhibited a negative association with a firm's stock market performance, no statistically significant association was found with return on equity. In the case of return on assets, board composition is negatively correlated with ROA, suggesting that board with higher proportion of non-executive independent members may lead to prolonged deliberations and decision-making delays, adversely affecting firm performance.

Regarding CEO duality, it was observed that Indian investors generally view it unfavourably. The presence of separate individuals at the helm of the board and management can prolong decision-making processes, potentially resulting in missed opportunities. As a result, investors tend to shy away from firms with CEO duality, preferring those with a single individual occupying both positions for streamlined decision-making and prompt opportunity utilization. Conversely, a positive association between the presence of additional independent directors in the audit committee and firm performance, encompassing both stock market and accounting metrics is observed. This suggests that augmenting the number of independent directors enhances organizational vigilance, promoting greater financial transparency and mitigating opportunities for earnings management, ultimately facilitating accurate accounting disclosures.

Research Implications

The composition of audit committee is clearly coming out as a factor, which investors look carefully while investing in companies. This factor is also having a positive effect on the accounting performance of the company. Therefore, a company wanting to have positive image in the market should have more independent directors in its audit committee. The independent directors in the board and CEO duality are not favourably considered by investors, mostly because of delay in business decisions which would have a negative impact on the performance of the firm. Although various corporate governance reform committees have recommended for two different persons at these two highest positions, however, the apprehensions of investors should be addressed carefully.

Limitations of the Study and Scope of Further Research

This study has few limitations. It was limited to BSE 200 companies of the Bombay Stock Exchange and the final sample came down to 123 companies because of insufficiency of data related to some of the companies and for not considering the banking and Non-banking financial companies. A larger number of companies could have been considered as sample to represent the whole population of listed companies and to have more universal findings. We have considered only 13 years of financial data for the study. In this study we have considered only a few governance variables; more factors could have been taken. A lot of data has been extracted from the CMIE database, where the financial data is standardized to do a comparison of all the companies, those restrictions will also have some impact on the results of the study. In the future, all the above factors can be taken into account while doing research in the field of corporate governance.

It is not that easy to arrive at a universal conclusion because characteristics like the independence of a board/committee member is a highly relative terms. The true independence of directors and their active involvement in board decisions cannot be adequately assessed solely through quantitative data. Moreover, the informal power of the CEO, including factors like family ties and participation in board sub-committees, can significantly influence company decisions and subsequently impact firm performance. Thus, forthcoming research should enquire into the detailed functions of the board by incorporating qualitative information to uncover hidden associations between board characteristics and firm performance. Similarly, in the context of audit committees, future studies can focus on evaluating the integrity and impartiality of committee members, as well as the quality of discussions in audit committee meetings and reports. This approach would offer a more comprehensive insight into the intricacies of the corporate governance mechanisms and their implications for company performance.

References

Abidin, Z. Z., Kamal, N. M., and Jusoff, K. (2009). Board structure and corporate performance in Malaysia. *International Journal of Economics and Finance*, 1(1), 150–164.

Alqatamin, R. M. (2018). Audit committee effectiveness and company performance: evidence from Jordan. *Accounting and Finance Research*, 7(2), 48.

Anderson, C. A., and Anthony, R. N. (1986). The Corporate Director. New York: Wiley Publication.

Bansal, N., and Sharma, A. K. (2016). Audit committee, corporate governance and firm performance: empirical evidence from India. *International Journal of Economics and Finance*, 8(3), 103.

Baysinger, B. D., and Butler, H. N. (1985). The role of corporate law in the theory of the firm. *The Journal of Law and Economics*, 28(1), 179–191.

Beasley, M. S. (1996). An empirical analysis of the relation between the board of director composition and financial statement fraud. *Accounting Review*, 71(4), 443–465.

Berg, S. V., and Smith, S. K. (1978). CEO and board chairman: A quantitative study of dual vs. unitary board leadership. *Directors and Boards*, 3(1), 34–39.

Bhagat, S., and Black, B. (2001). The non-correlation between board independence and long-term firm performance. *Journal of Corporation Law*, 27, 231.

Bijalwan, J. G., and Madan, P. (2013). Board composition, ownership structure and firm performance. *Research Journal of Economics and Business Studies*, 2(6), 86–101.

Boyd, B. K. (1995). CEO duality and firm performance: a contingency model. *Strategic Management Journal*, 16(4), 301–312.

Cadbury Report (1992). http://en.wikipedia.org/wiki/Cadbury_Report.

Carcello, J. V., and Neal, T. L. (2003). Audit committee characteristics and auditor dismissals following 'new' going-concern reports. *The Accounting Review*, 78(1), 95–117.

Carter, David, Frank P. D'Souza, Betty J. Simkins, and W. Gary Simpson. (2007). The diversity of corporate board committees and firm financial performance. *Available at SSRN 972763*.

Daily, Catherine M., and Dan R. Dalton. (1992). The relationship between governance structure and corporate performance in entrepreneurial firms. *Journal of Business Venturing* 7(5), 375–386.

Dechow, P. M., Sloan, R. G., and Sweeney, A. P. (1996). Causes and consequences of earnings manipulation: an analysis of firms subject to enforcement actions by the SEC. *Contemporary Accounting Research*, 13(1), 1–36.

Donaldson, L., and Davis, J. H. (1991). Stewardship theory or agency theory: CEO governance and shareholder returns. *Australian Journal of Management*, 16(1), 49–64.

Fama, E. F., and Jensen, M. C. (1983a). Separation of ownership and control. *Journal of Law and Economics*, 26, 301–325.

Fama, E. F., and Jensen, M. C. (1983b). Separation of ownership and control. *The Journal of Law and Economics*, 26(2), 301–325.

Fariha, R., Hossain, M. M., and Ghosh, R. (2022). Board characteristics, audit committee attributes and firm performance: empirical evidence from emerging economy. *Asian Journal of Accounting Research*, 7(1), 84–96.

Fosberg, Richard H. (1989). Outside directors and managerial monitoring. *Akron Business and Economic Review* 20(2), 24–32.

Gaur, A. S., and Gaur, S. S. (2006). Statistical Methods for Practice and Research: A Guide to Data Analysis using SPSS. Sage.

Goel, U., Bansal, N., and Sharma, A. K. (2015). Impact of corporate governance practices on working capital management efficiency: a structural equation modelling approach. *Indian Journal of Finance*, 9(1), 38–48.

Hermalin, B. E., and Weisbach, M. S. (1991). The effects of board composition and direct incentives on firm performance. *Financial Management*, 20(4), 101–112.

Kaldor, N. (1966). Marginal productivity and the macro-economic theories of distribution: comment on Samuelson and Modigliani. *The Review of Economic Studies*, 33(4), 309–319.

Klein, A. (1998). Firm performance and board committee structure. *The Journal of Law and Economics*, 41(1), 275–304.

Klein, P., Shapiro, D., and Young, J. (2004). Board independence and the family-owned firm. *Canadian Investment Review*, 17(3), 8–16.

Klein, P., Shapiro, D., and Young, J. (2005). Corporate governance, family ownership and firm value: the Canadian evidence. *Corporate Governance: An International Review*, 13(6), 769–784.

Kota, H. B., and Tomar, S. (2010). Corporate governance practices in Indian firms. *Journal of Management and Organization*, 16(2), 266–279.

Kumar, Naresh and Sudesh. (2019). Does corporate governance affect bank performance? Empirical evidence from India. *Prabandhan: Indian Journal of Management* 12(3), 7–23.

Lam, T. Y., and Lee, S. K. (2008). CEO duality and firm performance: evidence from Hong Kong. *Corporate Governance: The International Journal of Business in Society*, 8(3), 299–316.

McMullen, Dorothy A., and Kannan Raghunandan. (1996). Enhancing audit committee effectiveness. *Journal of Accountancy* 182(2), 79.

Mohd Saleh, Norman, Takiah Mohd Iskandar, and Mohd Mohid Rahmat. (2007). Audit committee characteristics and earnings management: Evidence from Malaysia. *Asian Review of Accounting* 15(2), 147–163.

Molz, Rick. (1988). Managerial domination of boards of directors and financial performance. *Journal of Business Research* 16(3), 235–249.

Nicholson, G. J., and Kiel, G. C. (2004). A framework for diagnosing board effectiveness. *Corporate Governance: An International Review*, 12(4), 442–460.

Persons, O. S. (2005). The relation between the new corporate governance rules and the likelihood of financial statement fraud. *Review of Accounting and Finance*, 4(2), 125–148.

Rechner, P. L., and Dalton, D. R. (1991). CEO duality and organizational performance: a longitudinal analysis. *Strategic Management Journal*, 12(2), 155–160.

Reddy, M. A., and Akula, R. (2011). Corporate governance-the role of various stakeholders. *Prabandhan: Indian Journal of Management*, 4(5), 50–55.

Routray, S. K., and Bal, R. K. (2016). Board composition, board gender diversity and firm performance: evidence from India. *Parikalpana: KIIT Journal of Management*, 12(2), 107–120.

Shivdasani, A., and Yermack, D. (1999). CEO involvement in the selection of new board members: an empirical analysis. *The Journal of Finance*, 54(5), 1829–1853.

Saat, Nur Ashikin Mohd, Yusuf Karbhari, Saed Heravi, and Annuar Md Nassir. (2011). Effective oversight roles of board of directors–the case of listed firms on Bursa Malaysia. *World review of business research*. 1(1), 231–245.

Vafeas, N., and Theodorou, E. (1998). The relationship between board structure and firm performance in the UK. *The British Accounting Review*, 30(4), 383–407.

Venkatraman, K., and Selvam, M. (2014). Impact of corporate governance practices and firm performance: an empirical study. *Indian Journal of Finance*, 8(12), 34–45.

Weir, C., and Laing, D. (2001). Governance structures, director independence and corporate performance in the UK. *European Business Review*, 13(2), 86–95.

Wijethilake, C., and Ekanayake, A. (2020). CEO duality and firm performance: the moderating roles of CEO informal power and board involvements. *Social Responsibility Journal*, 16(8), 1453–1474.

Wild, J. J. (1994). Managerial accountability to shareholders: audit committees and the explanatory power of earnings for returns. *The British Accounting Review*, 26(4), 353–374.

For Product Safety Concerns and Information please contact our EU
representative GPSR@taylorandfrancis.com
Taylor & Francis Verlag GmbH, Kaufingerstraße 24, 80331 München, Germany

www.ingramcontent.com/pod-product-compliance
Ingram Content Group UK Ltd.
Pitfield, Milton Keynes, MK11 3LW, UK
UKHW050234200926
475702UK00020B/3032
9781032916835